COVER and ENDPAPERS: Dr. Huckleberry and Garibaldi, mid-1920s. View out over mill and Tillamook Bay. Family's first home (and goat, see p. 26) are circled inside front cover.

"This county is strung out along the beach like a shoestring. Some places there is a quarter-mile of level country, some places the hills stick out into the water so that a shelf has to be dug out to build a road on. Also the rivers have a little level bottom land."

View south from Neahkahnie Mountain, at northwestern boundary of Tillamook County, over Nehalem Bay. Note man and horse on slope at left.
(Gifford Collection, OHS)

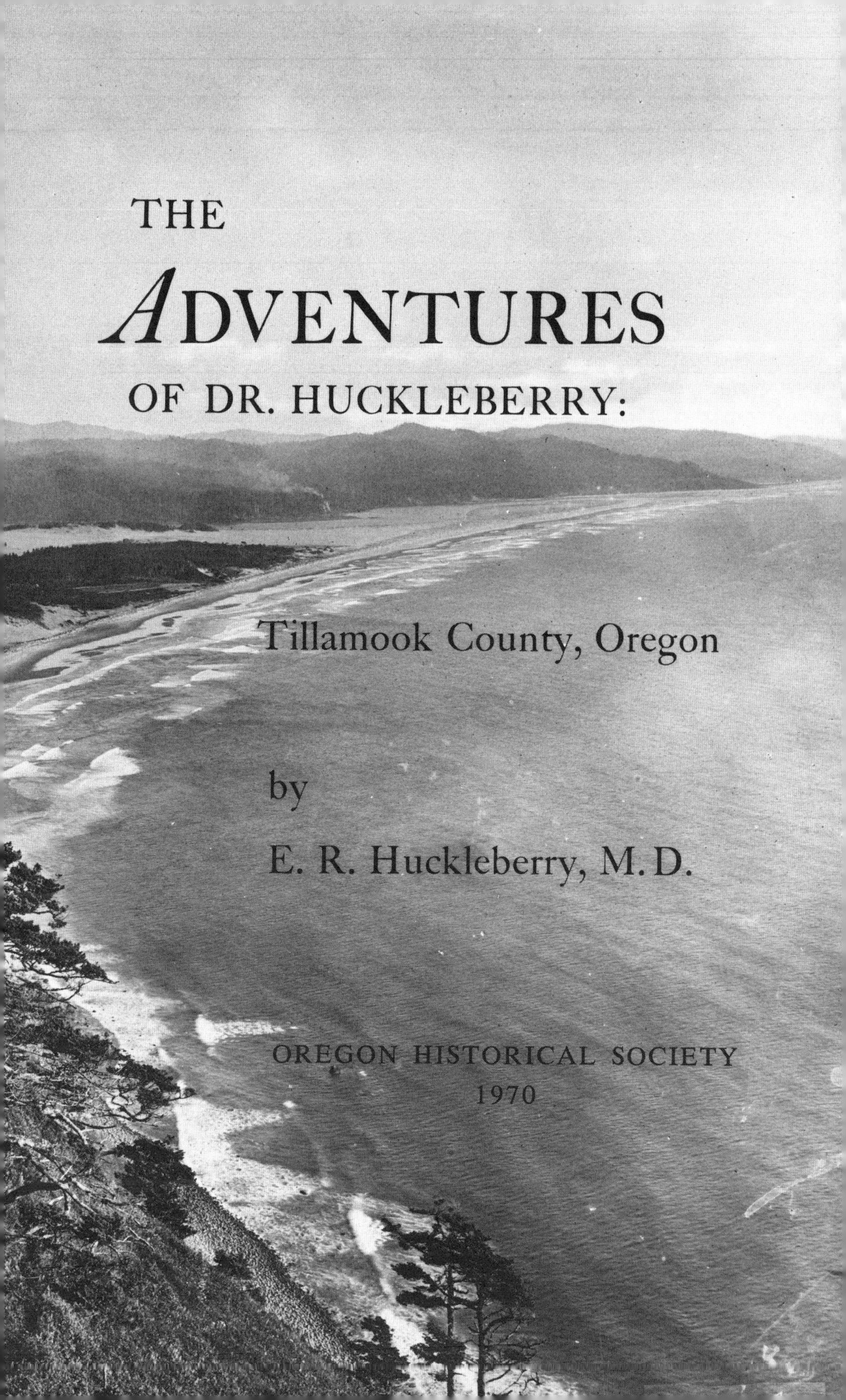

THE *A*DVENTURES OF DR. HUCKLEBERRY:

Tillamook County, Oregon

by

E. R. Huckleberry, M.D.

OREGON HISTORICAL SOCIETY
1970

LC No. 70-129303
SBN 87595-025-6

Printed in U.S.A.
Sweeney, Krist, & Dimm
Portland, Oregon

Preface

This little volume was written because I wanted to leave some record of a way of life that has disappeared. It was a good life, in spite of its limitations.

It is a factual account. I had first-hand knowledge of most of the incidents, and all others were told me by people I believe to be reliable. In many cases, names and locales are changed, for obvious reasons. I kept no diary or other record, so all the incidents described were dredged up from memory. For some reason I do not quite understand, most of the things I remember are things we laughed about; some of them because they were funny (at least to us at the time), and some because we couldn't do anything else about them.

In presenting this to you, I am under deep obligations to many people. There were those who urged me to compile it, mostly those who wanted to preserve some record of the time and place. There were those who participated in the events, who freely shared their experiences. Many of these people are no longer with us, the last few years having taken a heavy toll. I am grateful for the interest and understanding of the people at the Oregon Historical Society. Last, but not least, there is my wife, who, though a city girl, lived the life with me, who corrected my spelling and diction, and who typed the manuscript. To all, my sincerest thanks.

Recalling these happenings has been a great pleasure. It brought to mind many fine people and interesting events. I was happy in the situation. I am glad I had the experience. I regret that so many present-day physicians do not have such an opportunity to participate in the lives and aspirations of their patients.

The practice of medicine has changed radically. Doctors now know more than I did, have better training, better drugs, better equipment, finer skills. Even the language has changed. But I am repeatedly surprised and pleased by the deference and respect shown by some younger colleagues. I think maybe they vaguely sense that they have missed something important. If this account of medical practice as it was can help them to understand the development of their profession, it will be worthwhile. And it may help non-medical readers to appreciate what is now available to them.

E. R. HUCKLEBERRY, M.D.

Contents

Illustrations

THE ADVENTURES OF
DR. HUCKLEBERRY:
TILLAMOOK COUNTY, OREGON

The old Hudson and I. Ready to go! (Author's collection)

Tillamook Beginnings

One day John Jorgensen came into my office, sat down heavily in the patient's chair and said:

"Doc, I'm sick."

"How are you sick?"

"I dunno, I'm just sick."

"Where do you hurt?"

"Don't hurt nowhere, I'm just sick."

"Sleep good last night?"

"Yep."

"Eat breakfast O.K.?"

"Yep."

"Bowels O.K.?"

"Yep."

"Been hit by anything on the job?"

"Nope."

"Horse kick you?"

"Nope."

"Wife mad at you?"

"Nope."

"Well, where are you sick?"

"I dunno, I'm just sick."

"How do you know you're sick?"

"My eatin' tobakker don't taste good. I been achawin' Star for mor'n fifty year, ever since I was twelve years old, an' this is the first time it don't taste good. I gotta be sick."

He was, too. He was coming down with pneumonia, and for a few days it was anybody's guess as to whether he would pull through, but he finally made it.

I never saw anywhere in the books that a change in the taste of Star tobacco is one of the points in the differential diagnosis of broncho-pneumonia, but it is. At least it was!

That was just a typical incident in the life of a doctor in the big woods. Every patient could be, and often was, a unique prob-

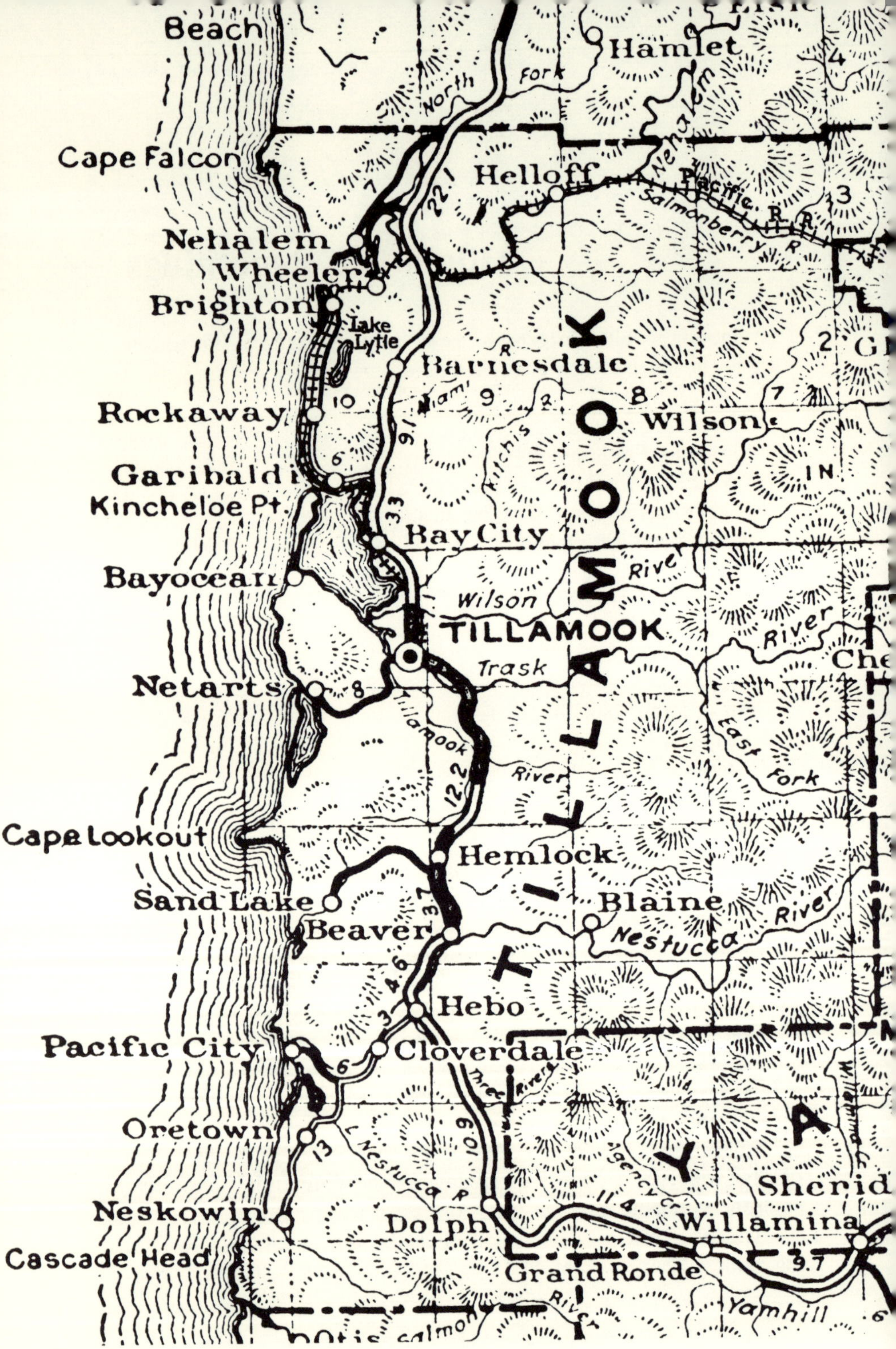

Tillamook County in the 1920s, showing roads and railroad.
Only parts of roads were paved.

lem. No two days were alike in the Tillamook country in the twenties.

I had come to Tillamook, it might be said, by stages and disposition. When my classroom education at Rush Medical College ended in 1921, I was accepted as an interne at Los Angeles County Hospital. Internes in that institution then got the munificent stipend of ten dollars a month, and all they could eat. I was not an outstanding interne, but I was a good eater. Because of the illness of my mother, we went to Portland, Oregon, on the completion of my interneship.

I was worse than broke, with a wife and baby, so grabbed the first remunerative job that came along. This was with the Oregon State Board of Health, doing educational work in venereal disease. I was provided with a projector and a few films. I went all over the state, giving a show and lecture, wherever I could gather up a few people to look and listen. Most of the time was spent away from the towns, putting on programs in logging camps, country schools or churches, teachers' institutes, empty store buildings, even in pool halls and taverns. Transportation was by passenger train when possible, but more often on a logging train, freight truck, handcar, wagon or anything that happened to be going my way. The friendliness of those frontier people, their willingness to help, and the interest they displayed in what must have been a pretty crude performance, was a constant source of surprise. In the course of these wanderings into the byways, I had a chance to see the beautiful timber lands of the western part of the state, and the mountains and high country of the eastern part. Oregon is a big state, nearly 97,000 square miles. It is about 300 miles north and south, 400 miles east and west.

The Cascade Mountains cross the state, dividing it into eastern and western parts. East of the Cascades it is high country, relatively dry, with many mountains, pine timber, mining and range livestock. There was some agriculture along streams and a few irrigation projects were under way. Where it was not too rough, large tracts were used to raise dry land wheat. Some of this wheat land was so steep, no eastern farmer would have considered farming it. He would have thought it impossible to get harvesting machinery over it.

West of the Cascades is the broad, well-watered Willamette

River Valley. At that time, much of the Willamette Valley had been logged off and converted to agriculture (grains, livestock and fruit), though there was still a great deal of timber on the slopes on both sides of the valley. Farther south, the timber of the Rogue River Valley was hardly touched. The wild and rugged beauty of the Rogue and its tributaries has been recorded by more competent pens than mine.

The western boundary of both the Willamette and Rogue valleys is the Coast Range. This is a low, north and south range, in some places several miles from the ocean, at others coming close. In 1922, from the clearings in the Willamette Valley, across the Coast Range, and down to the ocean, most of the country was covered with virgin timber. The little valleys, where logging was easy, and the land fertile, had been cleared, but the hills carried heavy stands of large Douglas fir, spruce, hemlock and cedar. These trees were so large, and stood so close together, that they permanently shaded the ground, choking off nearly all vegetation, except moss. The stillness, the vistas between the mammoth trunks, suggested a great cathedral. Where there was a break in the shade, the underbrush grew so thick a rabbit couldn't get through unless he carried a brush hook!

When the appropriation for the venereal disease program was used up, and my barnstorming through the backwoods was ended, I was contacted by Dr. Robert T. Boals, of Tillamook. He had x-ray burns on his hands and some fingers had just been removed. He needed someone to serve as his hands.

Tillamook is a small town, the county seat of Tillamook County, directly west of Portland. The name Tillamook is derived from the name of the Indian tribe in that area, and is thought to mean "meeting place of many waters." There are seven rivers in the county, some of them of quite respectable size. Five of them empty into Tillamook Bay, a good harbor, deep enough for medium sized ocean-going boats, but entered over a treacherous bar. To the south of Tillamook Bay is Nestucca Bay, and to the north, Nehalem Bay, both able to take small vessels. Tillamook County is about 50 miles north and south, and more than half that wide. It is on the western slope of the Coast Range, with a narrow alluvial plain along the coast and running up the rivers. When I first knew it, this plain was largely cleared and used for agriculture, but much of the remainder was covered with virgin timber.

The first and later white settlers came to Tillamook County by boat or walking over the mountains. The area was hemmed in, Coast Range on the east, with spurs extending west to the ocean, marking both northern and southern limits, so it was difficult to travel to or from the county. It is understandable that the people came to speak of "coming in" to the county, and that any trip to the surrounding areas was "going out" or "going outside." This idiom prevails even today, though there are good roads in all directions. People "go out" to Portland or to The (Willamette) Valley.

Because of the profuse rainfall (about 100 inches per year) and the long growing season (due to the moderating influence of the Japanese Current), grass grows abundantly, and that suggested cattle. The uncertain transportation (pack horses over the mountains, or the small coastwise boats, both very subject to weather condition) made it necessary to find a product that was not bulky, not too perishable, and readily salable. All this added up to cheese, and by the time I came on the scene, cheese was the principal agricultural product. It had started with each farmer making cheese from the milk of his own cows; later, groups of farmers went together and put up little neighborhood cheese factories. In the early 1900s, a chap named Carl Haberlach organized these little factories into a cooperative, with standardized methods of making cheese, quality control, and centralized selling. Under his leadership, Tillamook Cheese became the leading brand on the Pacific Coast.

The logging industry, which had moved steadily westward from New England, "letting daylight into the woods," had arrived in the Pacific Northwest many years before. There were many large sawmills along the Willamette and Columbia rivers, with camps back in the hills to furnish them logs. The lumber moved east and south by the trunk railroads that converged on Portland, and by boat to California and the Orient. When a branch railroad was completed to the Tillamook country in 1911, it was possible to get its timber to the domestic market, so the loggers moved in. In 1923, logging and sawmilling was the largest source of income for the county. There were five large, or fair sized mills, and several

small ones, with supporting camps. The coming of the railroad also made the Tillamook beaches accessible, and by 1923, a considerable number of people, mostly from the Portland area, were coming in. Accommodations to care for these people were beginning to be built, some reasonably comfortable, most of them not so good.

In spite of the fact that it was the twentieth century, conditions were quite primitive over most of the county. The main streets of Tillamook were paved, and for a couple of miles north and south, there were pavements built shortly before. Aside from that the roads were terrible. The best of them were made by laying 3 x 12 planks longitudinally on log cross-ties, one plank for each wheel track. Of course, if you got off the planks you were in trouble, but you soon learned to stay on! "You'll be all right if you just drive straight and keep your chaw in the middle of your mouth," one oldtimer told me.

The mouths of all the streams were obstructed by great heaps of brush and logs, washed down through the centuries. Caught by these obstructions were great quantities of twigs, leaves, and silt, forming dams, around and through which were intricate channels by which the water could get past. These were known locally as "rack heaps," and extended miles up the rivers and their tributaries. The channels kept open through and around these rack heaps were just enough to take the normal flow of the streams, and when one of the frequent heavy rains increased the flow, the low land was flooded. It was nothing unusual to see a rowboat in a farmer's back yard, tied to the porch rail. It was a good bet he would need it after the next heavy rain! These rack heaps have since been cleared away, and where necessary, dikes built, so floods are not so frequent, nor so widespread.

The county's "original" population, of course, was Indian, but in 1923 there were only a few pure blood Indians around. They were usually short, rather heavy-set, and tended to take on weight as they got older. They were quite dark. Some people claimed there was a trace of Negro blood in them, derived from a few Negroes among the sailors shipwrecked in the early days. The Coast Indians at one time practiced flattening the heads by binding a board on the heads of tiny babies, but I saw only two or three examples in very old people. I never saw a baby with its head bound. The pure bloods and half breeds usually kept pretty much to themselves, though some of the women married white men. They lived in little villages near Garibaldi and Nehalem. The men were fishermen for the most part, but they would do other work between fishing seasons. Some of them, particularly the younger ones, learned to work in the woods, and they were good workers. I never knew one to have a garden or keep any livestock except for dogs or a few chickens. Their only handicraft seemed to be weaving baskets and making articles pertaining to fishing. They built boats, wove nets, made the incidental accessories to fishing, such as oars, net floats, crab traps, etc.

The housekeeping chores of the women were minimal so they had time for other activities. When the fish runs were heavy, they would help their men work the nets. The pure blood women made baskets of grasses and cat-tail leaves which they gathered, dried, and sometimes dyed. Some of these baskets werc beautifully made, with small strands of closely woven material, mostly natural color, decorated with an occasional red or blue strand. I am told that some of these would hold water, but all I ever saw showed very small holes when held up to the light. Maybe when they were wet the fibers swelled up to plug the leaks, as happens to a wooden boat. They also made coarser baskets they called work baskets, most of them fairly large. They used them when digging clams, fishing, gathering driftwood, or for other rough or dirty jobs. The younger generation did not seem interested in baskets; they thought they were oldfashioned; bags and sacks from the store were easier to get, etc. When the older generation of women died, the art of making baskets died with them.

Another activity of the women was digging clams, which were plentiful on all the mud flats exposed at low tide.

Hogs are pretty good land clearers. If confined to a small area, they will soon root out and kill all weeds and brush. If there are any skunk cabbages, they will follow the roots so far that all that is visible above ground is a happily twitching tail.

One chap was clearing a tract near Lake Lytle in that way. One of his sows had a nice litter of pigs. When they were about three weeks old, he began to miss them; there was one less each day. Search of the lot did not turn up any dead pigs, or any evidence the other pigs had eaten them. He could not find a hole under the fence or any evidence that a bear had climbed over. So he hid in the bushes and watched. After a time, an old Indian appeared, climbed over the fence, caught a little pig, choked it to stop the squeal, and started off. The man stopped him and asked why he was stealing the pigs. "Why not—good eating," was all the reply he received. There didn't seem to be any good answer to that. Indians had been free to eat anything they could catch for no one knows how many generations.

There was a bulb that looked something like an onion which the Indians dried for winter use. They called it "wookus." Do not know how it tasted, but it didn't look good.

By the time I came on the scene, the Indians wore cloth clothing. Most of the men wore blue belt overalls, nondescript shirts and felt hats, usually black. Fishermen wore rubber boots, at least while working about the water. Women wore most any kind of loose fitting, ankle length dresses, and large, men's style shoes.

The younger generation went to school with the white kids and seemed to get along with them. Many of them dropped out of school as soon as they could get a job; the more energetic drifted out of Tillamook County, and I don't know what happened to them. Those who stayed were good workers when they worked—

that is, when they were hungry—but their improvidence and natural weakness for firewater kept many of them on marginal subsistence.

During World War II a number of the young men from the county were in the service and some married overseas, English, German or French girls. Sometimes that led to trouble!

Among certain classes in Europe, there was, and probably still is, the impression that all Americans are rich. In the American armed services, white, red, yellow or black are on equal footing; all are Americans. Therefore, all are rich. Some of those girls that married Indian Americans with the idea that they would live in luxury the rest of their lives, had a terrible shock. When they saw the homes their husbands came from, and the houses they were expected to live in, it was so far from their dreams that they rebelled. In some cases it was necessary for the authorities to step in and try to work out a solution. For a few, transportation back to their country of origin was provided. Some wanted to stay in this country and were assisted in getting training so they might be self supporting. A few that had salable skills, mostly in the needlework trades, as I got the story, were helped to find a place to use those skills.

Some of these foreign brides accepted life as they found it and did the best they could with the cards life dealt them. My recollection is that included most of them. Those boys that had learned a trade in the service, or took advantage of GI educational opportunities found a place for themselves in the American society, and, I believe, had no more and no less, domestic problems than their white buddies.

Of the girls married to boys that went back to the reservation, some of them, probably more than half, seemed satisfied with their lot. My impression was that they had come from very poor circumstances in the old country, and conditions we would consider very primitive were satisfactory, or even luxurious, for them. The various public and private agencics recognized the problem and made a special effort to help. Health, law enforcement, relief agencies and private organizations cooperated to assist these girls make the adjustment. In general, the results were very good. They made an effort to make a home for their husbands and children. Their kids went to the public school, seemed to average out about as well as the other kids. They raised a minimal amount of hell as teenagers, later settled down and became responsible citizens.

The small minority that didn't turn out well caused considerable trouble and attracted attention. It is always the spectacular minority that people know about; we tend to forget the majority that go along quietly and handle their own problems.

Looking back on it, my conclusion is that the marriage of European girls to Tillamook Indian Americans turned out fairly well. Probably they could be compared with marriage to any of the minority groups. Since there were few of these in the Tillamook area, I have no basis for judgment.

Living with the Indians in Squawtown, when we went to Garibaldi, was a Negro woman, quite a spry little old gal, whom I guessed to be in the late fifties or early sixties. One day I asked her how old she was. "Lan' sakes, white boy, I don't know how old I is. All I know is I was a good big gal, maybe 16, 17 year old when de nigger stealers ketched me in Tennessee and took me to Texas."

Figure that one out. I sure goofed on it. Never did find out how she got from Texas to Oregon, or why she was living with the Indians.

Sam Whitehorse was rather old, but he could still handle a net. One year the weather was unusually cold for that country, and on the last day of the commercial fishing season there was a thin rim of ice along the Nehalem River. Old Sam's net was set near the shore; part of his cork line was frozen into the ice. He made no effort to loosen it, just kept on picking out any fish that got caught. The local fish warden lived near Sam, liked him and didn't want any trouble with him, so tried to persuade him to take in his net, but no luck.

"Can't get out. Froze in," was all the answer he could get. Things went on like that for three days, during which time old Sam was getting a lot of fish, for there was nobody else to compete with him.

The warden explained his dilemma to the sheriff and they cooked up a scheme to throw a scare into Sam. Next morning, the sheriff and three deputies showed up at Sam's float. Each had a revolver prominent on each hip, and each carried a shotgun. Sam was out in his boat, running his net. The sheriff ordered him to shore, and sternly told him they had come to take him to jail if that net wasn't out of the water in thirty minutes. Sam took a quick look at all the artillery, then at the faces of the men who were frowning to keep from laughing, and said, "Get 'em out damn quick," and the net was in his boat in less than the specified time. Onlookers said Sam hadn't moved so fast since his squaw ran him out of the house with a bucket of boiling water.

In 1923, Tillamook was a town of possibly 2,500 people, and maybe there were 3,500 more in the county. There were five doctors in Tillamook City, and I made six, but one older man retired a few months after I arrived there. There were two out in smaller towns.

My wife and baby stayed in Portland when I went to Tillamook, planning to come on when I was able to find housing. The day I went in, it started raining, and in the next two weeks we had 40 inches of precipitation. Bridges on roads and railroad were washed out, telephone and telegraph lines were down, and for nearly three weeks our family had no communication.

During that time, I acted as hands for Dr. Boals, and had a chance to get to know him well. He was a unique individual, a short, square man, getting rather fat, with an abrupt manner, a stock of profanity that was the envy of every logger (and was the best argument for the value of a college education that ever hit Tillamook County). He had an underlying kindness of heart which he tried his best to conceal under his gruff exterior. He had been on his own since he was twelve, when he started as waterboy in a logging camp; then became a semi-pro pugilist. I never did find out why and how he got into medicine. Nobody could have been more kind to a green, tenderfoot medic, and I will always remember him with affection. I never had, or expect to have, a more rewarding relationship with any man.

Doc Boals in a characteristic stance.
(Courtesy Mrs. Mildred Brand)

Doctor Boals

One night we got a maternity call a couple of miles out of town. We picked up the prospective grandmother, who lived in town, and started out. Most of the road was under water, but not too deep for the old Buick, if no culverts were washed out. Dr. Boals put on a pair of hip boots (which I soon found out to be as much a part of the necessary equipment of a car as a spare tire), and walked ahead in the beam of the headlights. After a while, the old lady in the back seat leaned forward, tapped me on the shoulder and said, "Young feller, how long you been in this country?"

"Three days."

Then she looked at Dr. Boals, splashing along ahead of us and said, "Ole Doc Boals, he's been froggin' around in these here parts for nigh onto thirty year."

I thought "frogging" was the correct verb!

When we finished this case, the husband set out a lunch of bread, cheese and coffee. I was astounded to see him cut the cheese in chunks two or three inches square. In the Midwest, where I was brought up, cheese had been a luxury to be eaten sparingly and slowly. I found out later that one could buy a five-pound brick of Tillamook's best for 75 cents.

Another thing that surprised me was that many of these people ate cheese for breakfast; still another dietary custom new to me was the use of cream and sugar on baked beans.

A couple of weeks later, when the flood had subsided, we had a maternity call several miles out of town. The house was back from the road, possibly a quarter of a mile, surrounded by trees. We were afraid to drive up the hill to it because of the likelihood of getting stuck in the mud. So we left the car on the road and carried our equipment to the house. It was still light, and we had no trouble going in, but by the time we were ready to leave, it was darker than the inside of a cow's stomach. We lost the trail and blundered into a hole made by a windfall tree.Of course it was full of water left by the flood, and it was cold—January!

A story told of Dr. Boals, and one that could have happened, whether or not it ever did, involved a trip to Sand Lake. That was an area in the south end of the county, cut off by a very rugged range of hills. We had an urgent call down there about the middle of an afternoon of a very stormy day. He drove as far as his Buick could negotiate the road, then borrowed a team, buckboard and saddle from a farmer. When it was impossible to take a wheeled vehicle further, he put the saddle on one horse and went on. When the horse could go no further, he tied him to a tree and went on afoot, lugging the well-worn black bag. It was morning when he reached his destination, and the family was at breakfast. They invited him to eat with them, but he said, in his abrupt way, "Where's the patient?"

They indicated a fat teenage girl, sitting at the table stowing away a big plate of hotcakes. Turning to the father, Dr. Boals said, "My fee is twenty dollars."

The man was surprised, but dug up a twenty dollar gold piece. Boals pocketed it and turned to leave.

"Aren't you even going to examine her?" the mother asked.

"No. Give her another plate of hotcakes and let her go to hell."

And he stalked out and started the long trip home.

Another story about him was told me by the fellow involved. Bill Arnold was a logger, a moose of a man, rather short, but wide, weighing probably 225 pounds, with not an ounce of extra fat. He was what was locally called a "skookum (powerful) boy." For some reason, I do not remember what, Bill and Dr. Boals had a disagreement about an account. Bill had a good reputation for meeting his obligations, but this account he flatly refused to pay. One day Dr. Boals was looking out of the window of his second story office and saw Bill on the street. He opened the window and called, "Hey, Bill, I want to see you."

Bill climbed the stairs and went into the waiting room. Dr. Boals met him, asked him into the examining room, closed the door, and said, "Bill, when are you going to pay me what you owe me?"

"Never!"

Dr. Boals locked the door, threw the key out of the window, and said, "I'm going to take it out of your hide."

When Bill told me the story, he chuckled and said, "I saw the little banty rooster meant it, so I paid him."

That is one way to improve the collection percentage. Might be a good one but I never tried it—afraid some debtor might not have Bill's sense of humor!

One day a crew were rebuilding a bridge that had washed out. They had just gotten the stringers in place when another freshet carried away the falsework, including the temporary walkway. One man picked that time to break his leg, and to make it worse, he chose the far side of the bridge to do it. When Dr. Boals drove up, the workmen had rigged a splint on the leg, and were trying to decide how to get him across the river. The water was too swift and rough for a boat, and nobody was willing to try carrying a stretcher across on the stringers. Dr. Boals solved the question by getting the man on his back and walking a six-inch stringer across the river! Just to walk across was quite a feat; to carry a man larger than himself didn't make it any easier. The fellow who told me this story didn't mention the comfort of the man carried.

An incident quite characteristic of Dr. Boals came after we had been working together a few years. I had decided to take some time off for post-graduate work and had secured a Dr. McCannell to fill in for me. He had recently migrated to Oregon from North Dakota, seeking a milder climate. He had good references, though neither of us knew him. About a week after I got to San Francisco I received a letter from Dr. Boals in which he said: "Dr. Mac is a good surgeon, but he is a damn bull-headed Scotsman." They were too much alike to get along, though each respected the ability of the other.

Soon after getting to Tillamook County, I found there was a medical feud going on, as was true in too many small communities. This dated back many years, and the original cause I never knew, but as Dr. Boals was the leader of one faction, I was automatically involved, though I tried to be on good terms with all the other doctors in the community. Before many months had passed, a man of the other faction was killed in an automobile accident. One of his friends suggested that it would be a gesture appreciated by the deceased's family if all the doctors of the county would attend the funeral in a group. Next time I saw Dr. Boals I asked him if he was going.

View of Dr. Boals' hospital after addition was made to front, about the time Dr. Huckleberry was there. (Courtesy Mrs. Mildred Brand)

"Nope!" was the emphatic reply. "I try to live by the Golden Rule."

Rather puzzled by the statement, I asked what he meant.

"If I was dead, I wouldn't want that old so and so at my funeral. I'll treat him like I'd want him to treat me. I'll stay away."

I never was exactly sure what I thought of that incident.

Our hospital, which Dr. Boals owned, contained about 20 beds, and was in a big old dwelling house. There were two stories, so patients for the second floor, who could not get there under their own steam, had to be carried up, and post-operatives had to be carried down. As we said it: "We took them up and down on the back elevator." There was an x-ray, rather unusual in that country at that time, and a laboratory where the usual tests could be made. The operating room was small, but well stocked with instruments. Later Dr. Boals built an addition to the hospital, almost doubling the number of beds, and including an elevator.

Dr. Boals would tackle anything, which was fortunate, for there was nobody available that was any better, and both the road and railroad were too rough and slow and subject to interruptions, to ship out an emergency. On the whole, I thought the surgery he did was equal to that I had seen at Los Angeles County Hospital. Of course we got in over our heads at times, but I've seen that happen in big hospitals, too. The only difference was that we couldn't holler for help. Dr. Boals was most ingenious about wiggling out of a tight spot. We sort of got the idea that there was no emergency we couldn't handle. I assisted him in the operating room for several years, until the x-ray burns on his hands forced him out of surgery, and he left the county.

One afternoon an old man was brought to the hospital by three of his sons. He was choking and coughing. He declared he had swallowed his false teeth, the lowers, and they were lodged in his throat. The eldest son was skeptical.

"The old man is a little off his rocker and gets funny ideas at times. I don't know whether he swallered them teeth or not, but it is just like him to pertend. He puts on a wing-ding like this about every whip-stitch."

The old gentleman was hopping mad. He insisted he could feel those teeth still in his throat. Between coughing spells he cussed his sons and all their families for numbskulls and idiots. There was a real flap around that little hospital for a time. An x-ray film showed nothing, but a local dentist, consulted by phone, said some dentures were not radio-opaque. We tried to look down his throat, but nobody was very good with the laryngeal mirror, and the old fellow was not at all cooperative. In the midst of the uproar one of the daughters-in-law came charging in, waving a lower denture.

"I told you that old goat is crazy, I found this under his pillow."

This set the old gentleman off on another paroxysm of choking and coughing, but as soon as he could speak, he started to swear at this daughter-in-law, calling her the dumbest of a dumb family, declaring why his idiot son was so double damned crazy as to marry her was beyond all comprehension. "Them is my Sunday teeth, stupid; the ones I swallered is my everyday teeth. This is Tuesday, ain't it?"

In spite of this the family insisted that the old coot was looney and some day would have to be put away. All this conversation did nothing to allay the temper of the old man. The madder he got, the more he tried to talk, and the more he tried to talk, the more he choked. Finally, after a particularly violent spell of coughing, he vomited copiously, and with that, out came the teeth. He retrieved the denture from the mess on the floor; holding it above his head he danced around the room, yelling incoherently. Then he started shoving the teeth under the nose of each member of the family in turn, demanding: "Who is looney now? Who is looney now?"

One of the patients at Dr. Boals' hospital was there for care of cardiorenal disease. Condition was serious. He drifted in and out of consciousness; sometimes it was hard to tell just what state he was in. In some of his unconscious periods, he was quite restless, so to keep him from falling out of bed, we had put up side boards. In his conscious periods, he resented those boards passionately and repeatedly asked that they be removed, but it was not thought safe to do so.

One day while making rounds, I paused by his bed. He was quiet; his eyes were closed, and he was breathing quietly. I decided he was not conscious. Then his eyes opened a little, and he made a weak gesture that I should come closer. I leaned over him and in a barely audible voice he said: "Do something for me, Doc?"

"Sure. What is it?"

"Teach me to sing 'Don't Fence Me In.' "

Strange as it may seem, nursing help was no problem. We had a few permanent nurses, local women. For some reason I could not understand, there were a lot of drifting nurses in that country. They were mostly eastern girls, out to see the country, traveling two or three together, working long enough to get a stake, then moving on. They were good nurses, too, usually. How they could get that far off the beaten path was a mystery to me, but they did it. Maybe they were looking for husbands, for a few of them did marry local boys. We seldom had the same operating room crew two weeks in a row.

For a time we had a handyman about the hospital who had a remarkable nose for things to eat. Whenever the cook made pie, cookies, or any other goody, he was there. He made himself particularly obnoxious to the nurses when they were fixing themselves something special on their hours off. Hints rolled off his back; sarcastic remarks went right over his head. One day a couple of the girls made a batch of fudge. Sure enough, old Sam showed up just as they were cutting it into squares. He helped himself to a gen-

Dr. Boals' office was on the second floor of the IOOF building, in Tillamook, over Stillwell's store. The bank had the corner, first floor, and the Creamery Association the remainder of the second floor.

erous quantity and was busily stowing it away when one girl said to the other, "Mother's milk sure makes good fudge, doesn't it?" Old Sam looked startled, turned green, and hurriedly left the room, holding his hand over his mouth. Problem solved!

The cook lived in a little house on the street behind the hospital. One spring she got the idea she would raise some chickens, so she got a couple of hundred baby chicks. When these chicks were about two weeks old, there came an unseasonable cold snap, and one night the weather man predicted below freezing temperature. Cook hadn't counted on that kind of weather, and had no place to protect her pets. In desperation she decided to put them in the hospital kitchen after the day's work was done, planning to get them out before Dr. Boals came around next morning. She took some boards and made a pen behind the range (in which the fire never went out), and covered the floor with papers. She thought her babies would be as snug as moths in a buffalo robe. However, somehow, one of the boards comprising the pen fell over, and before the night nurse discovered what had happened, those chicks had scattered all over the first floor of the hospital. Of course they were frightened at the strange surroundings, and they were cold, so they set up a terrible racket. They hid under the beds and other furniture and nearly ran the night nurse crazy before she finally corralled all those chirping little monsters.

Garibaldi

When Dr. Boals' hands had improved enough that he could do his work, he offered me a salary for a year, to be followed by a partnership; or, as an alternative, there was a big sawmill being built at Garibaldi, twelve miles north (now only nine because of road straightening), where they were looking for a doctor. I could go there to establish a practice, have the use of his hospital, and help him with surgery, for which he would pay me. I elected to go on my own.

Care of industrial employees in that area was mostly on a contract basis at that time, and Dr. Boals had contracts with most of the mills and camps around Tillamook. After I had been at Garibaldi a short time, he and I were given the Garibaldi mill and camp contracts. He furnished the hospital facilities and looked after those hospitalized. I did the work at Garibaldi and in the camps, and helped with the surgery. Our contract covered everything that could legally happen to a man, except industrial accidents, which were the responsibility of the State Industrial Accident Commission. The contracts included medication and hospitalization. For this we received $1.00 per month per man. Later we received $1.50, and included industrial accidents. We divided the take fifty-fifty, and each furnished his own supplies.

There has been much criticism of contract practice. Probably some of it is justified. The important factor is the attitude of those holding the contracts. I am still convinced, however, that we gave the men as good service as was possible under the circumstances, and better than they could have obtained on their own. Certainly they got it for less money than similar services cost now. Some sort of arrangement was necessary to induce competent men to settle in the isolated communities. There is no doubt the psychological effect on the patient, offered limited or no choice, is bad, but possibly control of the quality of the service he gets is some compensation. I have rather a dim view of the ability of the ordinary person to judge the quality of medical service. If the patient gets well, he thinks the care was good; if results are not what he thinks they should be, it was bad.

There were many stories floating around about how Garibaldi got its name. One story has it that a shipwrecked Italian sailor was washed ashore there. Another is that the name is a corruption of an old Indian name. McArthur's *Oregon Geographic Names* reports that its first postmaster was an admirer of the Great Liberator.

The first recorded history of the Tillamook country is in the log of the good ship *Lady Washington,* Captain Robert Gray, master. In 1788 he sailed into the bay, thinking he had discovered "The Great River of the West." He stayed a few days, long enough to take on a supply of wood and water, and to do enough exploring to convince him he had not found the Great River. He traded with the Indians for fish, venison, furs, berries, etc., but his stay was terminated by an argument with the Indians, causing him to name the place "Murderers' Bay." He sailed north but missed the Columbia River, which he discovered four years later, on a subsequent voyage, and named it for the ship he was then sailing, the *Columbia.* Captain Gray reported Indian villages about the bay. Because of the geography, some of them were probably at, or near, the present site of Garibaldi. In 1923, there was a small settlement of Indians nearby, called "Squawtown," possibly a remnant of an older settlement. In early settlers' journals there are references to a settlement at Garibaldi, mostly fishermen. It was a location convenient to the best fishing grounds in the bay.

The mill was located there because of a number of factors. Among them were rail transportation, extensive sheltered tide flats suitable for log storage pens, an area of level ground large enough for the mill and lumber storage yard, adjacent area suitable for residences, and the availability of salt water transportation. To attract a permanent crew, the company had laid out a townsite and sold lots to employees for a low price. They also allowed the men to buy lumber, to be paid for by small monthly payroll deductions. The power plant was built large enough to supply juice to the town, as well as the mill. When we went there, the only completed houses belonged to some of the original fishermen or Indian families, but all over the place people were building their own homes. Some threw up a temporary shed to live in; some built a

Aerial view of Garibaldi and Tillamook Bay entrance at left. Around 1940. (Courtesy Oregon State Highway Division)

garage and lived in that; some lived in tent houses. In some cases, the house was built right over the tent. Lights were burning and there was the sound of saws and hammers until long after midnight. Most of the men were not carpenters, but they were building a home; and with the help of the women, and the advice and help of neighbors and friends, freely given, they got houses built, most of which are still standing.

I was almost broke when I went to Tillamook, but when we went to Garibaldi I had what we had saved from my wages, probably one or two hundred dollars. I had a stethoscope from medical school days (which I used until retirement from practice). A surgical supply house in Portland gave me a fifty dollar credit, on which I got a blood pressure instrument, an army surplus examining table, a pair of obstetrical forceps, two or three artery forceps, a Kelly pad, a few bandages, etc. A piece of razor blade clamped in an artery forceps made a usable scalpel. The local Ford dealer sold me a Model T roadster, complete with top and windshield, filled it with gas and oil, and took my note for the whole thing, as I remember, about $350.

An office was something of a problem, but there was an old pool hall, dating back to the fishing village days, that was not operating. The equipment was still in it, but I rented the room for $10 a month, with the proviso I would not move out the equipment. By moving the two tables (I never realized before how heavy a pool table is!) to the back of the room, I managed to clear about fifteen feet of the front. Some lumber from the mill made a partition between waiting room and examining room (where the pool tables were), some benches, shelves, etc., were added, and I was in business.

Living quarters were something else again. We found some rooms in one of the old houses, in which were a wood heating stove and cook stove. We had previously acquired a bed and baby's crib, a few dishes, some pots and pans. The remainder of the necessary furniture I made. Fuel was to be had for picking up driftwood on the beach, and the Ford roadster brought up many a load. We were able to buy scraps from the sawmill and later sometimes got a load of wood on a bill.

Getting started was a slow job. The first month I had one cash customer, $1.25. I remember his name, his complaint, his money.

The last I heard of him he was in jail, but to me he is still a good guy. I probably would have quit at the end of the first month, but didn't have the money to get away. Next month the book showed $87 collected, and when it went over $100 the third month, I thought I really had the world by the tail.

Food available was about what you would expect. Dairy products were plentiful; fish were to be had for the taking; clams and crabs could be picked up on the beach; illegal venison was rather common; bear meat was frequently for sale in the butcher shop. The grocery store usually had most of the ordinary items.

One factor that was present much of the time was mud. Most of the soil was decayed vegetation that churned up on the slightest disturbance. Until we got a lot of crushed rock on the roads, they were difficult. Cars stuck in the middle of the road were common. I remember seeing one in so deep the fenders were out of sight. Never did know how they got it out, or what the inside looked like.

We had a telephone, party line. There were only 10 phones on our line (less than the average), and all 10 phones rang on every call. My office ring was a long and two shorts; home was a long and four shorts. It was only a short time until I never heard any others but would wake up every time for either of my rings. The country exchange had its advantages, too. If we wanted to locate someone, we just told Bennie, the operator, who had the switchboard in her living room, and she usually knew where he was or could soon find him. If we wanted to go someplace, we would tell her, and the results were better than most of the modern high priced answering services.

Adjustment to this sort of life was probably harder for my wife than for me. I had lived in the country and in small towns all my life until I left home to go to college, while she was born and raised in Chicago. However, she stuck with it and never complained, which I appreciated greatly. One episode of her adjustment came when she called me at the hospital to tell me, "Mrs. Curry chopped her boy in the head with a two bit ax." When I got to the Curry home, the story was that Mrs. Curry had been chopping wood with a double-bitted ax when the little boy ran up behind her and was hit by the backswing.

I soon found out that the people in that area expected a doctor to be a fast driver, although what was fast with those cars and

A Tillamook County road of those days. (Gifford, OHS)

roads would be very moderate now. Forty miles an hour was really moving and that was possible on only a few short stretches of road. Many of the legends that grew up about Dr. Boals concerned his driving. A reputation for speed was good, legitimate publicity. People who would condemn speed in any other driver took pride in how fast a doctor could get over the road, especially if he was their doctor. I suppose it was considered a kind of guarantee of prompt service when needed; a sort of insurance they did not have to pay for. Stories of quick trips were told and retold, never losing anything in the telling.

When I went to Garibaldi, there was an elderly doctor at Bay City, four miles away, who had been practicing there probably 20 years or more. He was rather deliberate in everything he did, including his response to house calls. Many is the time, after answering a call to a family I had never served before, I was told, "We always used to call the Old Doc, but we were in a hurry this time, and we knew you would get here quick." There was nothing in that attitude to discourage a young fellow that liked to get up and move! One of the big thrills of my life came when I was able to plough past the old doctor stuck in a mud hole. I went by with the motor wide open, too fast to get stuck. Later, I heard he remarked, "I never saw a Ford roadster with wings before."

In 1926 or 27, an enterprising photographer climbed the hill back of Garibaldi and took a panorama picture of the town, then arranged with one of the White boys, a lad maybe 12 years old, to sell them. When the boy got to me I was not in a buying mood, so turned him down.

"It shows the whole town."

"No."

"It shows the mill."

"No."

"It shows your house."

"No."

"It shows your garage."

"No."

"It shows your car."

"No."

"It shows your goat." (My boy's pet.)

By that time my sales resistance was pretty well worn down. I thought that any kid who had studied his merchandise that well deserved a sale, so I bought it, still have it and now treasure it.

I have been told there are no poisonous snakes west of the Coast Range. At least I never heard of any in Tillamook County. However, there are great numbers of what are locally called slugs. These are like snails without shells, two to four inches long, as big around as a man's thumb, sticky, slimy creatures, that leave slime trails every place they crawl. As far as I know, they are harmless, except to tender plants, but most disgusting. They live in damp places and die if they get dry. We moved into a house in Garibaldi that had a great many trees around it and the shrubbery had not been pruned for a long time. The result was a damp property, ideal for slugs, and they were there in great profusion. I gave my boy, then about eight years old, a contract to gather them up and dispose of them at a penny apiece. By the end of the second day I owed him ten dollars and there was no noticeable reduction in the slug population. That looked like an expensive way to get rid of slugs, so I cancelled the contract. A small flock of ducks, four as I remember, did the job a lot cheaper. Those ducks would eat slugs until their front ends were so heavy they could hardly walk. They would go a few steps, when their breasts would come down on the ground. Then they would have to rest awhile before going a few more steps. They soon cleaned out the slugs, and ruthless pruning of the shrubs and trees made the place less attractive to the pests. The ducks were no longer needed, and the first thought was a series of duck dinners, but the more we thought about how they had become so fat, the less desirable they seemed, so I sold them to the butcher. He told me his customers were well pleased with the nice fat ducks. I didn't tell him what the ducks were fattened on, or ask who bought them. I had a conscience!

One of the pests of that country is a little gnat we called "No-see-ums." They are very tiny insects, present only at times, but then in enormous numbers. They get in your eyes, your nose and your mouth, and they have a pretty good bite for so small a bug. We had no insect repellants in those days, but were willing to try anything. Some claimed they would not bite a person who was well pickled in alcohol. I've seen it tried, but it seemed to me the 'picklee' attracted as many bugs—he just didn't care! And it did impair his working ability.

I think the best remedy was a liberal application of salty bacon grease, applied to all exposed body surfaces, and a bacon rind was often carried in the summer. Of course the salt was hard on any cut or sore, but that was better than the bugs. The grease caught any dust that came along, and by night a fellow would be pretty black.

John Langley had a string of horses. One morning he hunted me up to tell me that one of his horses had been kicked during the night, getting quite a cut, and that the Tillamook veterinarian was out and not expected back for many hours. Could I sew up the wound? I told him I'd never sewed on a horse, but if he could persuade the animal to hold still, I was sure I could do it. He agreed to that, so we went to his corral. The wound was a dilly, 18 or 20 inches long and very deep, in the left hip. John threw and tied the horse and I sewed up the hole in his muscle and hide. Just as we finished, I got an emergency call to the best hotel on the beach. There I found an official of the Southern Pacific Railroad, acutely ill from overeating of overripe seafood. From the clothing and the luggage, he was apparently a wealthy man, but just then he was only another fat guy with a king-sized bellyache. We soon had him reasonably comfortable, and only then did I realize I had not taken time to wash, and saw that my hands, arms and clothing were smeared with horse blood. If he had noticed it, he didn't say anything. I think he was too sick to care.

There is no more helpless feeling than to be overtaken by a sudden illness when away from home. You do not know whom to call for help. I have had that experience and I know. We saw it happen many times. The situation is particularly bad if all members of a group get sick at the same time. Once I saw a party of six, all down the same evening. These cases of food poisoning were more frequent before mechanical refrigeration became common, and when there was less control of roadside eating places. Most people who go to the beach want seafood, and seafood spoils quickly. With the best intentions, there is always the possibility of spoiled food being served. Because I have seen this happen so many times, when my wife and I travel, we never both order the same thing, no matter how nice the restaurant looks. One of us will be able to take charge if the other eats into a case of acute food poisoning.

Twelve women of Garibaldi had a bridge club that met once a month at the homes of the members. They made quite a thing of it. On the same night, their husbands would get together, more informally, at another home, for a session of penny ante or something. Late one afternoon before club evening, Paul Morgan, the mill superintendent, came to the office. Something he had eaten for lunch was raising a ruckus in his innards and refused to stay put. As he had already gotten rid of the offending material, all he needed was to get over the aftereffects, so I gave him a sedative and told him to go home and go to bed.

"I can't do that, the boys are coming to my house to play poker," he said.

"That's all right. You go home and go to bed. I'll play host in your place. I know your house well enough to get by."

After Paul got home, Mrs. Morgan called to know if I thought she should stay home to take care of Paul, but I assured her all he needed was sleep and he had to do that for himself. If anything happened, I'd be there and would call her if she was needed, so she could just as well go to the party and have a good time with the girls. She agreed, and told me she had made some apple pies for the boys, and would leave them on the table in the kitchen, along with plates, cups, forks and spoons, and the makings of coffee. All I'd have to do was make the coffee and serve the pie. That didn't sound too complicated. What she didn't tell me was that her little girls had made a big batch of fudge that afternoon, which hadn't hardened, so they set it aside, hoping it would be hard by morning.

When the time came for refreshments, I found the things as Mrs. Morgan had told me, so put the coffee on the stove, cut the pies and put them on the plates. Then I discovered that thick, gooey, chocolate stuff, and decided Pauline had forgotten to tell me about it. Since there was so much of it, evidently she wanted it to be served liberally. So I served it liberally! If the boys thought chocolate was a strange topping for apple pie, they were polite enough not to say it, and they all cleaned up their plates. However, next morning was something else again! When those girls discovered their candy was gone, they were puckered, and when those girls were puckered, they didn't keep it to themselves. I had to give them a box of store-bought candy to square myself!

One problem I had, that I suppose is still present in many a doctor's office, was to get an accurate location of emergencies. Apparently the person calling knows where he is and assumes that you do too. One advantage we had was the telephone operators and the party line. A typical caller would ask to speak to me in a big hurry, though the message could just as well be given to the girl answering the phone. Usually she could transmit a message much more quickly than I could come to the phone. One such call was "Doc, there has been a terrible accident out in front of my house. Come out just as quick as you can"—and slam! went the receiver on the hook, no name, no location. That telephone system covered an area roughly 50 by 25 miles. What to do! The phone operator remembered the call, because she noticed the caller seemed excited. It came in on the — line. There were 12 phones on that line, serving 12 farm homes, and it covered 25 or 30 square miles. She said she would see what she could do. Some time later she called back and said, "I've called every phone on that line and none of them know anything about an accident, except one, and they don't answer. My guess is they are all out at the accident."

With that to go on, I headed out towards the non-answering phone, something over 25 miles away. Sure enough, in front of that house, two automobiles were piled up. I don't know how an automatic phone with its seven digits would have handled that emergency!

One evening I was driving along the crooked road between Garibaldi and Bay City and came up behind an old Ford Model A. The road was too crooked to pass, so I followed. The old clunk ahead rattled and banged along and the gal driving it was busily talking to her companion. Suddenly the right rear tire came off and rolled along parallel to the car for a time, then veered and went off into the canyon. I saw this clearly by the light of my headlights, but the driver of the car either did not know anything had happened or didn't care. Possibly she just thought her old car was getting more noisy! She drove blissfully on, that naked rim rattling along in the gravel.

A somewhat similar story was told by Leland Bester. He said he saw a car stopped by the road, evidently in trouble. He stopped to offer help and found a chap, much the worse for drink, stupidly looking at a rear wheel which had no tire. He was sure all the tires were in place when he started, and had no idea what had become of the missing one. He had noticed the car was riding a little rough. At first he thought it was because the road was in terrible shape, but finally thought it might be something wrong with the car and got out to look. He just didn't see what could have happened; he really didn't. He didn't think anybody would steal a tire from a moving car—it would be too dangerous. Leland suggested that they go look for the tire, so the chap got in with him and they started back on the road which the drunk said he had traveled. They drove more than two miles before they found the tire by the side of the road!

There was a very good reason why the last couple of miles had been a bit rough.

Logs and Lumber Mill

The Garibaldi mill was located on the north edge of the bay, where there was level ground for the mill and yards, which covered about 40 acres; extensive tide flats for log storage pond, about 50 acres; and where the dock could be built out to water deep enough to float a ship. The storage pond was outlined by rows of piling. Connecting them were small (two feet or less) logs or poles, so arranged that they could move up and down along the piling as the water level changed with the rise and fall of the tides. The ponds were divided into pockets by similar rows of piling and floating poles. Logs were brought to the pond by water or by rail, later by trucks. Logs which had been put in the water somewhere else in the bay or in the tributary rivers, were brought in flat rafts, masses of logs surrounded by boom sticks, poles fastened end to end by short chains. These rafts were pulled by small tugboats. Logs from other places came via the ocean as rafts, or on barges.

There were two kinds of ocean going rafts, the Benson (Robertson) and the Davis rafts. Because of water conditions, only Davis rafts were brought into Tillamook Bay.

The Benson rafts were made up in large cradles, tapered at either end. They tried to have long logs for this raft, 50 to 150 feet, and they were placed in the cradle by donkey engines, always breaking joints. There was a large chain from end to end through the middle of the mass of logs, with branch chains attached every few feet. These branch chains went around the raft, holding the logs in position. When complete, the raft would be 1,000 feet long, 50 feet wide, 35 feet high, and would draw about 30 feet of water. It would contain five to seven million board feet of timber. Due to the cigar shape, and the way they were tied together, these rafts could be towed from either end. If the prospects were for good weather, a deck load of shingles, box shook, or sawed lumber was sometimes lashed on. Most of these rafts were made up along the Columbia River and were for the California trade.

Davis rafts were for shallower water, and were made without

Tillamook Bay log raft, above. Man may be Dr. Max R. Charlton. (Courtesy Mrs. Charlton and Mrs. Mildred Brand) Below is cigar raft. Note size of chain. (Gifford Col., OHS)

a cradle. First a mat of logs was brought together on a bit of quiet water, the longest logs on the outside. This was tapered a little to the square ends, and might be 12 to 20 logs wide in the middle. These logs were laced together, figure eight fashion, with heavy cable, up to one and a half inches, drawn very tight with steam engines. On this mat was placed another layer of logs, one log narrower, and the whole thing laced together as before. When finished, this raft might have six or eight layers of logs, and the top would be five or six logs wide. These rafts varied a great deal in size depending on the depth of water where they were made and where they were going. The larger ones had about 600 logs, might contain something over a million board feet, were about 200 feet long and 80 feet at the widest point. It would take five or six miles of cable to tie such a raft together. Those coming into Tillamook Bay were smaller, averaging about 500,000 to a million board feet. These rafts were made at various places along the Oregon and Washington coasts. Those received in Tillamook Bay came from Siletz Bay.

Both types of rafts were designed for ocean towing and were a cheap way to move logs along the coast. As far as I know, they were not taken across the ocean. A few were caught by storms and broken up, but most of them made the trip safely.

The barges were either old hulls with all the interior torn out, or large floating boxes especially built for the purpose. They were filled with logs and towed like the rafts.

All logs were branded in the woods on one or both ends, sometimes by paint, but usually by an indented mark. The brand was raised on the face of a heavy hammer, and when the end of the log was struck with the hammer, the brand was impressed into the wood. Sometimes in rough weather, logs would escape from a flat raft or from a pond and float free. As long as the brand was visible, ownership was clear. Fishermen or others out on the water in boats would sometimes get a line on such a free log and tow it in for a reward. Others not so honest might saw off a few inches from the end of the log, and have an unbranded log they could sell. Log pirates were not unknown; under cover of darkness they would release logs from rafts or ponds, cut off the ends, and claim the logs as their own. A good log was worth a lot of money and the business was profitable if one didn't get caught.

Logs coming by rail were dumped into the water. An unloading machine would dump a whole carload at one time, making quite a spectacular splash. Later, trucked logs were treated the same way. No matter how they got there, logs were sorted by species and grade, then stored in the various pockets until needed. When needed, these logs would be floated and pushed from the pockets into the saw pocket next to the mill, from which they were pulled up an incline by an endless chain, to the deck of the mill.

On the deck they were looked over very carefully, to remove any foreign material. While the logs were being dragged from the woods, sizable rocks were sometimes embedded in cracks or in the bark. If the head saw should hit one of these it might easily knock out a piece of steel that would go flying through the air like a piece of shrapnel, or the saw might break in two. A high speed band saw breaking can have serious consequences. The rock itself might become a projectile. After the sniper had picked out any foreign material he could find, the log was rolled on the carriage, a platform on wheels, running back and forth on steel rails, past the saw. It was fastened to the carriage securely, or "dogged" by steel hooks, so it would not shift while passing the saw. After a run past the saw, it would be moved forward on the carriage for another cut. This action was controlled by one or more men who rode the carriage, called the setter and the dogger. In later mills, the log is advanced mechanically and the setter and dogger are out of a job.

The carriage was moved in a variety of ways. In some mills, especially short log mills, the carriage is controlled by a large piston, operated by steam or compressed air. This is called a shotgun drive, because it is so fast.

Most carriages are moved by a cable attached to one end, then given a few turns around a steel drum driven by a steam engine or electric motor, then over a block at the far end of the carriage track, and then to the other end of the carriage. By this arrangement the carriage could be pulled back and forth by reversing the powering drum.

A band saw is an endless ribbon of steel, 12 to 16 inches wide, passing over wheels above and below the log on the carriage, set to cut parallel to the rails. It has teeth on one side or both, depending on the construction of the mill and type of logs. It is the filer's

Left, above, branded log. Note undercut and sawcut, making plain annual rings. Below shows log in mill with first cut. (OHS Collections)

Garibaldi mill, above, in early 1920s, with burning waste at left on small island. (Courtesy G. B. Abdill) Photo below shows log pens, big stack of later mill days, with wigwam burner. Kincheloe Point, upper left, marks bay entrance. (Courtesy Harry Scovell)

job to keep the saw sharp and in good cutting condition, with the cutting edge a little wider and having a little more tension than the back. It must cut uniformly across the face of the teeth or the saw will not run true and the resulting cut will be crooked. The filer has to have a few saws in reserve, for no one can tell when a rock or hard knot will knock a piece out of a saw, or when the head sawyer will decide he needs a sharp saw. A good filer can make or break a mill.

The head sawyer's position was alongside the carriage, close to the band wheels, behind a heavy wall to guard him from flying chips. He had a number of levers, mostly foot manipulated, by which he controlled the carriage. He gave signals by hand to the setters when he wanted the log advanced, turned, etc. He could have used more hands and feet if he had had them! It was his responsibility to saw the log up to the best advantage, to make the sort of lumber the mill could sell. This required a lot of skill and judgment.

The Garibaldi mill could cut logs up to eight feet in diameter. Larger logs were split with dynamite while still in the water. The sawmill itself was among the largest ever built, though not the largest. It cut 250,000 feet of lumber every eight-hour shift. The first step in sawing was the head rig, or primary saw. The older mills used large circular saws, but this one had a band saw. The large slabs, or cants, moved from the head rig on powered rolls to resaws and edgers, where they were converted into boards or timbers. These moved on to an outside sorting table, or "green chain," where various types of lumber were segregated. Drying might be accomplished by piling in the open yard, each layer separated by a "sticker" to "air dry," or it might be done with artificial heat and forced air circulation in a kiln, for four or five days. Then came the planing mill, where it was surfaced to sizes and patterns, after which the lumber was ready to go out to the trade.

One of the features of this mill was that it was electrically operated. Each machine had its own motor. There were no dangerous drive shafts or belts. A steam generating plant, fueled by sawdust and planer shavings, not only generated enough power to operate the mill, but also to supply the town. Disposal of waste was a problem in any sawmill. Manufacture of small wooden articles, such as lath, took part of it, but there was still much for the big "wigwam" burners that were found at every mill. Burning of waste gradually decreased as more uses were found for this material. It is now usually cut up into chips to be made into paper or compressed wood products.

The green chain at Garibaldi was a platform waist-high, about 30 feet wide and possibly 200 feet long. Freshly sawed lumber was carried over it by conveyor chains. Men stood on the floor alongside it, pulling off the lumber, sorting it, placing the various grades and sizes in separate piles. When a pile was large enough, it was picked up by a "straddle" type carrier and taken to the storage yard, dry kiln or planing mill. A few men worked on the platform, keeping the lumber straight for those pulling it off.

One time the company hired an efficiency expert to survey the plant, and among his recommendations was one to move the steps to the green chain platform to another place he thought would be more convenient. A few days later, one of the men working on the platform wanted a drink of water, went to the place where he had been going up and down for months, and walked off on nothing. After I got his skinned places patched up, and was filling out the accident report, we came to the question, "What caused the accident?" "That damn deficiency expert," was the reply.

Baseball interested sawmill men just as it did everybody else. One year, when radios were just beginning to come into use, there was some discontent among the crew because work interfered with their listening to the World Series. Solution of the problem was involved, but effective. Bill Murphy, the head filer, could see his home from the window of his filing shack over the headrig. He had a radio at home. At the end of each inning, Mrs. Murphy would chalk up the score on a blackboard on their porch. Bill could read it with the aid of field glasses. He would go down to the head rig and give the score to the boys there. They would chalk it on a board, put it on the chains where each man in the mill could see it as it went by; then it was sent out along the green chain, duly noted by each man. Then the lumber carriers took it to the planing mill, etc. The scheme wasn't perfect, but it worked and kept everybody well informed on current history.

The company was hit by the Great Depression and some internal problems. The mill was closed down, the machinery pulled out, and the crew scattered. Years later, Paul Morgan, who had been superintendent, produced the following lament. It may be a bit lacking in technique, but the nostalgia expressed is felt by all the oldtimers.

Goodbye Sawmill

The old buildings stand so quiet and still
Time was they were a busy mill.
Paint's all gone, the roof is rusty
Look inside and everything is dusty.

Heard no more the bandsaw's whine
As it plowed thru logs line after line,
Fir and hemlock, cedar and spruce
Many of the men liked their snoose.

Remember the log pond you liked to see
When the train dumped logs, you shouted with glee.
With a clatter and bang and great big crash,
The logs hit the water with a mighty splash.

Pondmen, pondmen, you were a husky lot
Out in the open, whether cold or hot.
With steel calked boots, pole with a pike
Some fell in and they were a sight.

The bullchain was noisy as it clanked up the slip
Carried the logs held tight in its grip.
Big log, little log, short and long
Made good lumber if nothing went wrong.

Down on the carriage rolled the log with a bump
The sawyer with his nigger bar hit with a thump.
People knew by the shotgun's bark
The mill was running, got a good start.

Rolls began to roar and chains their rattle,
The crew started working after their prattle.
Machinery was moving, saws began to sing,
It was a good sawmill, winter and spring.

Time was the mill didn't care two hoots
About the union's jurisdictional dispute.
While the men decided if they should go
The A.F. of L. or the C.I.O.

Happy days were there as the men went to work,
They liked their jobs, only a few did shirk.
The fragrance of new lumber and fresh sawdust
Was a tonic to the crew and to some a must.

Millman, planerman, tallyman, loader
Took care of the lumber they knew they oughter.
Time was the hauling was done by hosses
Good ones and bad ones like some of the bosses.

Gone is the train, but the pond is still there
The croaking of the frogs is all you can hear.
No more is heard the deep throated whistle
For the yard is full of weeds and thistle.

Doesn't seem right, these things should be
No jobs now for you and me.
But the sun still sets, and up comes the dawn
It is the way of the World—time marches on.

The people who are involved with an industry are an interesting part of it. Those in Tillamook County were no exception. Most of them were hard working, honest individuals, interested in their jobs, their kids, their recreation, just like people everywhere. They had their faults, made mistakes, and sometimes displayed amazing virtues.

Old Frank Hobson was a good man, well along in the sixties, a quiet little fellow, hard working; a faithful member and treasurer of one of the local churches. He had a hernia which caused him much trouble. Many times I was called to reduce it and sometimes had difficulty doing so. To his questions about permanent relief, the only answer was surgical repair. This he always refused, stating he could not afford to lose that much time from his job. I never pressed the matter, for he was a very bad surgical risk and it was doubtful if he could survive the operation.

One day he walked into the office and announced that he wanted his hernia repaired. To my surprised questions as to why he had come to that decision at that particular time, he said, "I can't go on this way; it is getting worse and pretty soon I'll not be able to work. I have a straight life insurance policy I've been paying on since I was a young man. Another payment is due in about a month; I can't meet it and the policy will be cancelled.

I know from the way you have talked about it you think I can't stand the operation. I own my house, but I haven't a thing to leave the old woman to live on, so if I'm going to die, I want to do it before that insurance policy runs out."

He came through the operation very well, but had a peculiar post-operative course. We could not see any reason for it, but he just didn't improve. He didn't seem to want to get well, and in four or five days died. Shortly before he died, he called his wife and told her he had just remembered he had spent twenty-five cents of church money at the grocery store, and that she must be sure to repay it. It has always seemed to me that little old man had elements of greatness.

One night I got a hurry up call to see Hans Suter. Hans was Swiss, spoke very broken English. He had been a dairyman, but had turned the farm over to his son-in-law, moved to Garibaldi and got a job in the mill. He was probably in the late fifties, but was an active, energetic, wiry little fellow who was holding down a hard job. When I got there, he was lying on the bed, all doubled up, with his hands clasped over his abdomen, obviously in severe pain. He was saying over and over, "Oh, I vish I didn't made de joke. Oh how I vish I didn't made de joke."

His wife had no idea what he was talking about, or didn't know enough English to tell me—I am not sure which. After giving him something to relieve the pain, I was able to get him straightened out enough to examine his abdomen and to talk a bit. His explanation: "Just last week I say I go to Huckleberry and say sick, pain in belly. Maybe I get to lay in bed for a few days. Oh, I vish I didn't made de joke."

It was clearly an acute abdomen, so we took him to the hospital and opened him up. He had a perforated peptic ulcer, though he denied any ulcer symptoms. He had plenty of reason for that pain in his belly, and he got several days in bed!

Johnny Swenson was planer mill foreman. He had a tremendous voice, the loudest talker I ever knew. Maybe it was from years of giving instructions to his men over the deafening shriek of a battery of planers. He was also an ardent fisherman. A neighbor of his, Fred Strohm, was also an enthusiastic fisherman, but he was a quiet man and Johnny's loud talk irritated him. One day I met him on the street. Before I could say anything, he held up

Main plank street of Rockaway Beach, 1917, with vacationers arriving on train, left. (Wesley Andrews Collection, OHS)

a finger and said, "Sh! Hear that?" I listened but could hear nothing unusual. "What was it?" "Johnny Swenson just caught a steelhead in Nehalem Bay!" (Twenty miles away.)

Lem Parker managed a box factory at Garibaldi. One of the machines in the factory was a planer for smoothing the little pieces of board used in making wooden boxes. One day he was showing a new man how to feed this machine. "Always use one piece to push another into the machine. Don't do it this way," and he started to push a piece in with his fingers. The piece broke and his hand was jerked into the feed rolls of the planer. He was thoroughly familiar with the machine so was able to stop it before his hand reached the knives, but it was pretty badly squeezed and had two or three broken bones. Fortunately it was his left hand.

The State Industrial Accident Commission accident report blank had three parts on the same sheet, one for the doctor, one for the employer and one for the employee. As he represented the company, he filled out the employer's part, then filled out the employee's section. One question was "What would you suggest to prevent a similar accident?" He wrote, "Don't be so damn careless."

Tales of Tillamook Folks

One day, when it was too stormy to work, a group of us were sitting around the stove in the mill office. In the group was Mr. Lamb, president of the bank in Tillamook, probably the richest man in the county, but a man who could fit into any group. After a while the talk turned to automobiles, and somebody asked Mr. Lamb what kind of a car he drove. "None—I can't afford a car."

Everybody laughed, thinking he was joking. He became serious. "I mean it. I can't afford it." We were still skeptical, so he said: "Two or three years ago my wife got the idea she wanted a car and bent my ear about it so much I had to take notice. I coudn't see why she wanted a car. We live close to the grocery store and the church, about the only places she goes. It is just a nice walk for me to go to the bank, about all the exercise I get. But she kept at me, and finally I asked her why she wanted it. She said she would like for us to get out in the country on nice Sunday afternoons and she would like to take her friends for a drive once in a while. So I told her O.K. 'You look around and see what kind of a car you would like to have.' After a week or two she decided what she wanted and found out how much it would cost. She didn't pick the cheapest car available, nor was it the most expensive. I took that amount in bonds from my safety deposit box and gave them to her, telling her they were hers to use as she pleased. If she wanted to buy a car it was fine with me, but I had one request—wait three months. 'Do all the riding you want to, but get a taxi. Don't be stingy, go wherever you want to go and whenever you want to. Keep a record of what you spend and compare it with the interest these bonds will bring you.' She did it, and at the end of three months had spent about five dollars less than the interest on those bonds. She had no depreciation, no license fee, no taxes, no flat tires, no insurance, no garage bills, no looking for a parking place, no worries. The program looked good enough to continue. She started a savings account with that five dollars and each quarter has had a little money to add to it. There is now something over $50 in that account which she says she is keeping for something special. I don't know what. And she still has her bonds! I just plain can't afford a car, and I don't think some of you fellows can, either."

One day I made a call on a shiftless family some distance from town. It was noon time and they asked me to have dinner with them, but things were so dirty I declined. I guess the woman saw me eyeing the dishes for she hastened to assure me they were clean, they were washed by many waters. I found out later they called their dog "Many Waters" because he was given to them when a pup, before he was housebroken, by an Indian neighbor, and they thought he should have an Indian name. I don't know if he was their dishwasher, but I had my opinion. Anyhow, I never regretted postponing lunch until I got home.

Every Saturday night, Sam Slocum would have a few beers at the tavern before going to his bachelor shack. He would not be pie-eyed drunk, just pleasantly tipsy. One Saturday some boys broke into his cabin, painted footprints from the door to the opposite side of the room, up the wall, across the ceiling to the door, down the wall and out the door. Sam came in after dark, lighted a lamp, saw the footprints on the floor, followed them to the wall, then up, then overhead. He leaned over backwards so far following those tracks he lost his balance, fell over backwards, and banged his head hard on the floor. When he sobered up and realized what had happened, he was so ashamed he passed up the tavern for three weeks.

One family that was an object of curiosity lived in our town for a time. They lived in a tumbledown shack; had practically nothing in it. The man worked very little, yet they seemed to eat regularly, and they had a nice, nearly new, car. We wondered how they did it. One cold, rainy day, their little girl, just too young to go to school, was seen wandering around. She was crying and cold, so a neighbor asked her in, took off her wet coat and shoes and set her before the fire to get warm. When asked why she was out in the rain she said, "The man came to collect a payment on the car and mamma made me get out."

Some people did not take marital relations very seriously. One fellow was sent to jail for a few months. While he was away, his wife took up with another man. Maybe she needed a meal ticket.

When her husband got back home, he found another woman to his liking. All four of them moved in together, to reduce overhead, and they all seemed happy with the arrangement.

Hugh Benjamin was a very pompous man, a pillar of the local Methodist Church. One Sunday morning he was on his way to church, dressed in his best clothes. There had been a heavy rain the night before, but the day was clear and bright. He came to a street crossing where there were blocks of wood for stepping stones across the mud. He didn't know that the night before a car had been stuck at that intersection, and when it was finally freed it left quite a hole which the rain had filled with water. In the middle of this hole was floating a block of wood. It looked all right, just like the others in the crossing, but when Hugh stepped on it, he went to his knee in muddy water. It threw him so off balance that he fell full length in the muddy street. That was hard on pomposity—it is difficult to be dignified when one is lying in a mud hole!

We had one sheriff that didn't carry a gun. His favorite weapon was a pick handle and he carried a bundle of them in his car all the time. They seemed to inspire a lot of respect. I talked to one fellow caught with some illegal venison, who said, "That guy was staked out along the trail as I was bringing that deer out of the hills. The night was dark as the inside of a black bull, and I was almost against him before I saw him. When he told me to drop that deer and my gun, I did it, 'cause I knew what he was holding in his hand. If he had had a gun, I'd have made a run for it, taking a chance he'd miss me in the dark, but I knew the first step I'd take, that pick handle would part my hair and my scalp with it."

One night this same sheriff called me to the office to fix up a fisherman who had gotten too drunk to remember about that pick handle. He and a couple of his buddies were soused enough that they decided to fish some illegal nets, and when they were caught, thought it would be great fun to throw the sheriff in the river. They didn't get it done, and this chap happened to be closest when the pick handle came into action. After I had him bandaged up, he started trying to talk the sheriff out of jailing him on the illegal fishing charge. He quite eloquently recalled that we were at war, that war called for food; the president was asking everybody to produce, and if he was turned loose he would get in his boat, fish all night, produce more food and help win the war. The sheriff was hard nosed—the three of them had to stand trial for illegal fishing; he was no better than his buddies who were already in jail. Besides, drunk as he was, he would probably fall in and drown. As he was led off down the hall, he was berating the sheriff for being a bottleneck and impeding the war effort. "Bottleneck, bottleneck, bottleneck," as long as I could hear him.

One winter some of the boys around Nehalem got rather bored and decided it might be fun to start a new lodge. They rented a hall and built up an elaborate ritual with a lot of monkey business. I do not remember all their routine, probably never knew it, but do recall some of it. Their main source of fun was the initiation of new members, and the center of this was a goat. They built a frame, covered it with blankets reaching the floor, and placed an old cow's skull and horns on the front end for a head. It was set on four wheels, each a different size and mounted off center by a different amount. The finished structure was almost as big as a cow. The motive power was furnished by a short, husky member of the tribe, who could get inside the goat, by bending over a bit. When he pushed this contraption briskly across the floor, with all four wheels turning in a different rhythm, the guy perched on top got a really thrilling ride, if he could stay on.

Another part of the initiation featured a star, painted high on the wall at one end of the hall. After a long buildup, the candidate was told he would undergo a test of his ability to carry out a task calling for great concentration. Then he was told to walk the length of the hall, looking at that star, never taking his eyes off of it by so much as blinking. While he was engrossed in this task, one of the brothers would slip up and put a handful of marbles in his path.

The officer who was to give the password was seated on a low stool, and the candidate was so placed that he had to bend sideways to hear the whispered word. As he was concentrating on what was being said, another brother would empty a medicine dropper of ice water in his upper ear.

The high point of the initiation came with the passing of the secret grip. They had obtained an electrical device, operating on a battery, that would give a pretty good jolt. This gadget was hidden behind the station of one of the officers. From it a wire went under the floor to the president's station, where an outlet was installed. The president had a wire up one pants leg, under his shirt, down his right sleeve, connected to an electrode for the palm of his hand. In the preliminary work this electrode was pushed up under his sleeve. When the time came for revealing the secret grip, the president quietly plugged the wire in his pants leg into the outlet from the shocker, and shifted the electrode to his palm. The side of the electrode next to the president's hand was insulated, but the other side was bare. There was a long lecture on the hidden significance of the grip to the brotherhood. The grip itself was extremely complicated, and when it was properly accomplished, the officer near the shocking gadget was supposed to kick the switch that activated it. While the candidate was dancing around trying to get loose from that contact, the president was holding his hand firmly, solemnly assuring him he was experiencing the thrill that comes to a worthy brother when he completes the proper grip with a kindred soul, and urging him to prolong the experience.

One day the president made a little mistake. He got the electrode upside down. When the shock came he did the dancing; the candidate held firmly and the other officer was so amused by the president's antics, he forgot to turn off the juice. Some of the group insisted he forgot intentionally, for he had had a little trouble with the president a few days before. At any rate, the president, who had been the prime mover in organizing the lodge, sort of lost his enthusiasm, and the brotherhood soon fell apart.

Al Bostique, a French Canadian widower about 70 years old, was a vigorous, corpulent, public spirited chap who had a hand in nearly everything going on. He knew everybody and liked people, and enjoyed life immensely. One day he showed up with a beautiful case of shingles. Shingles, technically *herpes zoster,* is a very disagreeable complaint, for which we could do little, but which runs a pretty definite course. It is not fatal, but is so distressing it sometimes reminds one of the old adage about seasickness—"first you are afraid you are going to die, then you are afraid you won't." Al had a severe case, and he became progressively more despondent. About the time the condition was due to improve, he came in, full of bounce, his old self, happy as a clam at high tide. He started shouting in his high squeaky voice before he reached the top of the stairs. "Doc, Doc, I know how to cure the shingles. I know how to cure the shingles."

"How, Al?"

"Yesterday I felt terrible. I felt awful. Last night I went out with four of my grandsons. I got drunk as a skunk, went to a dance and danced every dance until the orchestra went home. Then we went to a roadhouse and drank beer until daylight. The boys couldn't go to work this morning, but I feel fine. That's the way to cure the shingles."

During the dust bowl time, most of the Okies went to California, but a few of them strayed up our way. They had no skills, so all the work they could get was common labor. Common labor around a sawmill was hard work, so they didn't stay long. Really, most of them were so under-nourished they were not able to do hard work.

One who stayed a few months was a lay preacher. He had occasion to come to my office a few times, and when I could, I liked to talk with him. He was an interesting fellow, ignorant as a billy goat, but intelligent, and he had a lot of unique ideas. One day the conversation turned to the Bible and I said something about the Revised Version. He became quite indignant. "I don't want to hear nothin' about no revised virgin. I want my Bible just like Jesus and Moses and Abraham had it." I tried to tell him a little about translations, but could not get through to him. I think he honestly believed English was the only language that had ever existed, and that the King James version was a permanent monument from the beginning of time. There was a book in the office that had a picture of a page of Hebrew. I showed it to him, told him that was what Jesus had. He looked at it a minute, slammed the book down and said, "You are a damn liar," and stalked out.

One evening, an excited Swiss farmer called, almost incoherent. I could not make out what was wrong, but it was evidently something important, so I drove out. His wife was lying on the floor of the kitchen; the kitchen equipment was scattered about the room, and the woman, the furniture, the floor, the walls, and even the ceiling were spattered with what looked like blood. The husband was so excited he was worse than useless, and the woman's mother was sputtering around like a hen with a flock of baby ducks. I did get out of the man that he had come in from the barn, found his wife in this condition and had called his mother-in-law, who lived nearby. The mother could not speak English. She said a lot but did not tell me anything.

One look at the woman and it was evident she could not have lost that much blood and still have such a good color. Next thought—it doesn't smell like blood. A quick check of the woman revealed no serious injury, just a goose egg on her head, but that she was pretty badly scared. After a little encouragement, she was able to get up and prove to herself that she was all in one piece. Then the story came out. She had been cooking a batch of cranberries in a new pressure cooker. She had filled the kettle pretty full and evidently a piece of a berry had boiled up into the safety valve, plugging it. Then the inevitable explosion had occurred. Something had conked her on the head and she went down and out. Probably she had been hit by the lid of the cooker, for we found that battered article behind some furniture at the side of the room. When she waked up and saw all that "blood" she was sure she was fatally injured. What a mess! Don't know how long it took to clean up. I went home before somebody could think of asking me to help.

In Tillamook, the dairy started early deliveries about five o'clock in the morning, which, in the winter time, was pitch dark. One of the leading citizens began missing milk. His order was two quarts a day, and very frequently there would be only one. For a time he just thought the dairyman might be short of milk and was cutting down so he would have some for everybody, but when he got his monthly bill he was charged for two quarts every day. When he protested to the dairyman, he was assured there was plenty of milk and two quarts were always left at his back door. That was different, so he got up as early as the milkman came around and listened just inside his back door until daylight. He had done it only once or twice when he heard the bottles rattle. He jerked open the door, and with his deer rifle in hand yelled, "Hands up!"

And there in the light streaming from the open door, with a bottle of milk in hand, was his next door neighbor and best friend!

How would you handle a situation like that? I don't know how it was handled in this case. That part of the story was suppressed.

One Tillamook undertaker that was before my time was still talked about. He was a big man, with long, full Santa Claus white whiskers. He drove his own hearse, dressed in a black suit, black stovepipe hat, white gloves and shiny black boots. His hearse had a lot of plated metal trim which was kept shining at all times. His team of black, spirited animals could be held down to a funeral pace only by dint of good driving. The outfit was very impressive.

This man kept a notebook in which he recorded the height of everybody in the county. He said he wanted to give service. When he was called to a home, he always knew what size coffin to take. He said if he ever had to go back to his store for a longer coffin, he would be disgraced. A professional man should always be prepared to meet the needs of his clientele. He had what he called a furniture store, but most of his stock was coffins, which were always kept on display in his window.

This fellow was quite a joker when off duty. One Fourth of July, the community put on a big celebration. One feature was a parade. He entered a float. When it came out of his barn, it consisted of a very large flatbed wagon. On the rear of this flatbed were 12 or 15 dogs, tied up rather short. Of course the noise of the crowd and the band caused the dogs to set up a terrific commotion, barking, howling and fighting. On the front of the wagon was a large sized sausage grinder, turned by a small boy. The undertaker was tending the machine, and every little bit would walk to the dogs, pretend to pick one up, walk forward and pretend to feed it into the grinder. Then he would reach into the tub in front of the machine, pick up a length of link sausage and toss it to the kids along the street.

I think our change in attitude towards financing is as significant of the changes we have seen as anything can be. When I was young, paying cash for everything was a supreme virtue. In many cases you got a little discount if you paid cash. It was something to be whispered about if a neighbor was in debt, or had not paid for his home. To buy a home, you had to put down a considerable portion of the price. That was why so many people rented their dwelling places. Credit to a farmer was not considered debt, for he paid up in full when he sold his crop. Not to do so was a disgrace. Most grocery stores extended thirty-day credit to good customers only. Extending temporary credit for ordinary purchases was a courtesy to a friend or evidence that the seller was keen to sell for one reason or another. The share crop system, where the tenants never got out of debt from one year to another, was looked

Main Street, Tillamook, Oregon, late 1930s. (SP Collection, OHS)

on as a form of slavery. (I'm still not sure it wasn't.)

But things have changed. According to published figures, the man that has clear title to his home, his car, his business, is the exception. In many cases, even the clothes on his back are not his. We all have charge accounts, credit cards, make installment payments as a matter of course.

Shortly after I went to Tillamook, a wealthy man was dickering with an elderly farmer for his real estate. They came to an agreement; the purchaser pulled out his checkbook and wrote a check for the entire purchase price. The farmer refused to take it; he wanted an interest bearing note, which the purchaser refused to make, and the deal fell through. The incident attracted much attention. The idea of refusing cash was something most of us had never thought of.

But things changed rapidly. It was only a few years later that I bought a new automobile. Nothing was said about terms of payment. When the salesman delivered the car, he pulled out a long document, pointed to the line at the bottom and said, "Sign here."

"What for?"

"It is your sales contract."

I had money in the bank to pay for the car, so wrote a check. He took it and gave me a bill of sale, but he stormed out of the office, mad.

There was a farmer in the Nehalem Valley who was a hard working, honest fellow, rather slow thinking, but with an outgoing personality that never saw a stranger. He was on a first name, kidding, joking basis with everybody. Whatever put the idea in his head was never known, but one year he decided to run for the legislature. Nobody took his candidacy very seriously except himself, but to our surprise, he was elected.

The evening after the session ended, I went into a restaurant and there he sat. He saw me come in and called me to sit with him. He was the same old Joe, loud-voiced, joking and laughing, finding something funny in every subject that came up. Finally the legislative session was mentioned, and still kidding, I asked him how much under-the-table money he had brought home. Immediately he became serious. I think it was the first time I ever saw him serious, and it was most impressive. I was afraid I had really offended him.

"Huck, I tell you for an honest fact, nobody offered me anything. Nobody so much as bought me a meal or gave me a cigar. That is the straight truth. It worried me. I'd heard so much about the corruption and bribery around that statehouse in Salem, that I expected it to pop up several times a day, but I never saw a bit of evidence that it was going on. Eventually I came to the conclusion I was just too dumb to recognize what must be all around me. I looked forward to the time when the session would be ended and I could go home and start pumping cows again. At least the cows were stupider than me.

"On the way home, I drove past the farm of an old friend. He used to run a big place near Rickreall, but he is getting along in years now, and has rented most of his land. He just milks a few cows to have something to do. It was ten or eleven o'clock when I drove by his house; his chores were all done and he was sitting under a tree in his front yard reading the paper. I stopped for a little visit, since he wasn't busy. We talked about this and that. Finally I told him of my worries about my intelligence. He was quiet for a little while, then he said, 'Joe, do you see my old bull out there in the pasture?'

"Yeah, I see 'im."

"Well, I have no idee how he does it, but he never seems to have any trouble telling which cow is in heat. I think maybe a feller looking for somebody to take a bribe is as smart as that old bull."

Then Joe laughed again in his usual boisterous way and said,

"Huck, that made me feel good." But he did not run again in the next election.

For several years, Rollie Watkins was our representative in the state legislature. In some of the elections he was unopposed. One such election year, a group of pranksters started a campaign to write in Andy Gump. The present generation may not know that name, but the oldsters will remember him as a well known comic strip character, with a long nose, no chin and a prominent Adam's apple. The campaign caught fire, but Rollie persuaded the court to order that Gump votes not be counted. Rollie got less than half the usual number of votes that year, so we thought Andy was elected but got a dirty deal from machine politics.

Hans and Gertie Schwartz had a large family, and it took all they could get together to feed them. Hans worked in the mill, picked up all the after hours jobs he could find, and spent all his spare time fishing. Gertie was a good manager, kept her house clean and always had something for her kids to eat. Their house was rather back in the woods, for they had picked out a lot on the extreme edge of the townsite, and they had no close neighbors. They had cleared out a garden spot back of the house and always had a good supply of fresh vegetables in season. They also raised a lot of cabbage to put down for kraut in the winter. On the edge of their little clearing was a sort of tunnel, formed by the roots of trees that had started to grow on an old log. The roots had followed down the sides of the log; the log had finally rotted away, but the younger trees with their roots remained. The tunnel was about four feet high where the big end of the log had been, tapering down to about two feet high at the other end. The Schwartz kids used it for a playhouse and enjoyed it so much Hans had left it when clearing his lot.

One day, late in the summer, Gertie looked out at the garden and there was a bear having a feast in the cabbage patch. She was not about to hold still while anything ruined her cabbages, so she let out a yell and started for that bear, grabbing up an ax that was by the back door. I guess that bear knew he was where he had no business to be, or maybe he was frightened by her screech. At any rate, he started to leave. That tunnel looked like a good place to hide, so in he went, but pretty soon got to the place where he couldn't go any further. When he tried to back out, he couldn't, for broken ends of roots caught in his fur, and he was trapped. Gertie kept chopping at him through the cracks between roots and finally killed him. When they got him out, most of the meat was good, but she had ruined the hide. She had literally chopped it to pieces.

One of the merchants in our town had a family of girls, and he wanted a boy so badly he could taste it. He talked about it every time there was an opportunity. Tommy Stanton was one of the young fellows that grew up in the town. He was a nice enough kid, though most people thought he was pretty far back in the line when the brains were passed out. Probably the fact that he stuttered had something to do with that. However, he finished high school, got a job in the mill, and held it. One day he persuaded a nice girl to marry him. In due time she presented him with a fine pair of twin boys. The next day I dropped into the merchant's store, and stopped to chew the fat a little bit. Naturally the subject of the twins came up, and that triggered the usual spiel about how he wanted a boy, this time ending up with "I'd give a thousand dollars to anyone who could tell me how to get twin boys."

Just then Tommy walked in. Thinking to have a bit of fun, I said, "Tommy, Mr. Ross says he will give a thousand dollars to anyone who will tell him how to get twin boys."

Tommy's eyes popped open and his jaw dropped. He didn't know there was that much money in one place outside of a bank. However he recovered quickly. "W-w-ell," he stuttered. "I-I-I-can t-t-tell him how I-I-I d-d-did it."

Bud Jones was an intelligent fellow, a foreman in the mill. One day he came in the office complaining of loss of weight, excessive thirst, and constant tiredness. A quick check showed that he was diabetic. This was before insulin was in general use, and all we could do was control diet. I tried to explain the basic principles; he got the idea quickly and started to set up a routine for himself. He listed no sugar in the coffee, no syrup on the hot cakes; in fact, no hot cakes; no jelly sandwiches in his lunch pail, again no sugar in the coffee, no pie or cake. For supper, no spaghetti, no hot biscuits, no jam, no puddings, again no sugar in the coffee. In the evenings, no candy, no soda pop, no cookies, etc. Then he paused for a few minutes while thinking it over, and said, "Doc, I've just raised hell with myself."

Logging

The principal timber trees of the county were:

1. Douglas Fir, sometimes called Oregon pine, used for general construction, strong and durable.

2. Spruce, light weight, not so strong, used where weight was an item, as boxes, boats, etc. In the earlier days airplane frames were made of spruce.

3. Hemlock, used for general construction, not so strong or durable as fir.

4. Larch, similar to hemlock.

5. Cedar, light, brittle, resistant to rot, used for shingles and siding.

6. Alder, fine grain, moderately hard, takes good finish, used for furniture.

Some of these trees were quite large. Trees 20 to 30 feet in circumference at the base were common, and many were 150 feet to the first branches.

Logging was (and is) a most interesting industry. The time of bull teams had passed in 1923, but some of the old skid roads were still around and many who remembered them were also around. Skid roads, for "skidding" or sliding logs to water, were made of small logs, 12 to 24 inches in diameter, laid parallel across the proposed road a few feet apart, and sunk into the ground until only a few inches of the top side were visible. These were notched out in the middle to make a guide for the logs skidded. Then the surfaces were liberally daubed with grease. This grease was any kind available, tallow from any animal—beef, horse, deer, even porcupine, mixed with fir pitch. The route of a skid road was so laid out that it was down hill all the way, and the lower end was in water deep enough to float big logs.

In those days, timber was plentiful, so only logs that could be gotten onto the skid road without too much trouble were cut. They were maneuvered onto the skids by a combination of animal strength and manpower. Manpower was applied via peavy and

screw jacks. A peavy is a hardwood pole, about two and a half inches thick and five feet long, with a steel spike in one end. About eight inches from the spike end, is hinged a curved steel bar, with a sharp hook on the free end. With the spike jabbed into a log, and the hook applied a few inches laterally, a rotary force could be applied. The peavy was also a pry bar, a digging implement, a pounding tool, and sometimes a lethal weapon. Few of the loggers of that day had ever heard of the science of physics, but they were experts in applying physical principles to move logs weighing up to several tons!

Skid Road

Once on the skid road, two or more logs, depending on size, would be hooked together in tandem, by short chains with a spike on each end. Such a train of logs was called a "turn." A team of oxen, ranging from four to twelve animals, was hitched to the front. The hitch was applied in such a way that there was a slight lift of the front end of the turn when the oxen pulled. With a great shouting, cracking of whips, bellowing of oxen, and profanity, the massive logs would start to slide. At first it was hard going. The laboring oxen, with their heads almost to the ground, would heave and strain, urged on by whip and voice. As momentum increased, it was easier, and when the occasional steeper pitches were reached, the team was forced to gallop at top speed to keep from being overrun. Any poor beast unable to keep out of the way had broken legs, or worse. If caught so that a turn went over him, he was ground to a bloody mess. The carcass was rolled aside, where it was promptly butchered by the squaws that were always around looking for such an accident.

Ahead of each turn was a boy (grease monkey), with a bucket of grease to daub up dry spots on the skid logs. In dry weather, the turn was followed by a man with a bucket of water and a gunny sack, to take care of the fires sometimes started by friction of turn logs against skid logs.

Oxen were used in the woods because they could pull more, and they pulled more steadily than horses. They could get around better in steep, rocky or brushy country, and were more adept at avoiding accidents. They were also easier to obtain and cheaper, which was important because of the heavy mortality.

Oregon coastal fir forest. People just did not believe how big the trees were. (OHS)

Villages sometimes grew up about the lower ends of skid roads, though the immediate vicinity was not a desirable building site because of the noise and dirt of the turns. Then, too, the type of conversation thought necessary to secure satisfactory exertion from the ox team was not pleasant to delicate ears. As a result, the buildings on each side of a skid road were occupied by those not too fastidious and who hoped to profit from workmen using the road, such as saloons, cheap boarding houses, cheap vaudeville, etc. Those frequenting the enterprises were the down-and-outers, drunks, bums, petty criminals, prostitutes, along with a sprinkling of men actually working or looking for work. It is said that Seattle and other towns along Puget Sound, that were started during the bull team era, had skid roads down their main streets.

From the conditions along a skid road, any place with similar conditions was called a skid road, later corrupted to "skid row." The "skid row" apparently has now a firm place in American idiom, and probably few people using the term have any idea of its origin.

Chutes

When the steam donkey and railroads came into the logging industry, the need for a skid road nearly disappeared. The steam engines had so much power they did not need the skid roads. However, a variation, called a slide, or a chute, was sometimes used to move logs from a high place to a lower one, usually into water. In this case, logs 14 or 15 inches in diameter, were laid parallel longitudinally, with a third log in the bottom, to make a trough or chute down which the logs slid by gravity. There was one of these operated by Charley Stone in 1923. It was clearly visible from Garibaldi. This slide was 1,500 feet long, and it was quite a sight to see a giant log come sliding down, to stop with a tremendous splash in the pond at the foot of the hill—total time for the trip, 20 seconds.

A county road serving the camp and a couple of dairy ranches, crossed the lower end of this chute. The engineer of the donkey that started the logs down the slide could not see this road, but a special whistle signal was arranged to be sounded just before a log was released, and great pains were taken to see that everybody in camp and on the ranches understood the importance of keeping

Left, Stone Logging Co. chute, down which logs slid to pond opening into Miami River. Mr. Stone sits on side log. (Courtesy C. F. Stone)
Right shows gravity chute construction. (OHS)

that chute clear when a log was coming down. One of the ranchers had a boy who was rather venturesome. One day he was approaching the chute as the signal was given. Thinking he had time to get across before the log arrived, he speeded up, but killed his engine in the chute. He jumped out just in time to see the log hit his car, tear it to pieces and scatter them all over the surrounding landscape.

"That's a hell of a way to move a stalled car," was his only comment as he started to walk home.

Along this chute was a trail, but a sign at the bottom strictly forbade the use of that trail during working hours. Sometimes in summer, when things were dry, a log would hang up on this slide, due to friction. The remedy was to dump a lot of water in the upper end of the chute, enough to wet the log and the chute, and to jar the log loose. One summer day, the owner of another logging operation, came over to see how the chute was working, for he was considering building one. He was dressed in his best clothes. He paid no attention to the sign, walked up the trail studying the construction of the chute, not noticing that a log was hung up above him. The workmen above did not see him. They dumped water in the top of the chute as usual; the log started sliding and came roaring down, splashing an avalanche of mud and water to either side. The unsuspecting observer was knocked down by the force of the water and soaked to the skin. The foreman had seen the accident and came tearing down the trail. When he saw no serious damage was done, he said, "Dammit, Tom, you ought to know better than to walk up that trail while we are working."

"I do, now," was the reply, as he dug mud out of his eyes and ears.

A sled that fitted the slide was made for pulling supplies up the hill by cable from the donkey at the top. One evening at quitting time, a couple of the boys decided it would be easier to ride that sled down than to walk. No amount of talk could change their minds—they insisted that if the empty sled would slide down, it would slide with them in it. They were tired and didn't want to walk down that hill, they wanted to ride. So they got in the sled; it was pushed into the chute, and down they went.

Next day at quitting time, they were asked if they wanted to ride down again.

"No," was the answer.

"Why? We couldn't talk you out of it last night."

"That sled talks better than you do." They walked down the hill thereafter.

There were five fairly large logging operations in Tillamook County and several smaller ones. The first step in a logging operation was the cruise. This was a survey to determine the amount and kind of timber on a given tract of land. Two men, the cruiser and his assistant, the compass man, would go across the tract at regular intervals. The compass man would hold a compass in his hand to make sure they kept on the agreed course, and checked the distance traveled by counting his steps. By experience he knew the length of his step and how many steps it would take to traverse a quarter, half, or whole mile. The cruiser followed the compass man, observing the trees in a strip of agreed width, noting the varieties of trees, estimating the board feet of lumber and its quality. An attempt was made to have the strips observed typical of the whole tract. By applying the figures obtained for a known part of the tract, an estimate of the timber on the whole tract could be figured. In addition, the cruiser made notes of any features that might affect logging, such as slopes, rock outcrops, canyons, streams, etc. It was remarkable how accurate these cruisers were. A good cruiser would regularly estimate within five percent of the timber taken off as shown by the scale sheet.

The next step was to lay out a general plan for logging the area, including right-of-way from existing railroad to the tract, location of the railroad in the tract, including sidings, etc. In the early days, logging was usually directly into the water, but by 1923 most timber along water was gone. Small streams were of no value as the logs were too big.

Another necessary step was the selection of a campsite and construction of a camp. For the very large operations, the camp buildings were sometimes so constructed that they could be loaded on flatcars and moved to a new location as one area was cut off. The camps were built to house only men, with a few family houses for the key men. Some camps had enough families to need a one-room school. The key building was the kitchen and mess hall, under the tyrannical rule of a cook, sometimes a woman, assisted by a second cook and flunkies, waitresses, and a bull cook. The

Whitney Co.'s Blue Star Camp, at head of incline on Kilchis River. Portable buildings are mostly bunkhouses. (Courtesy G. B. Abdill)

bull cook was usually an older man, who kept up the wood supply and did the other heavier chores about camp. Near the kitchen and mess hall would be the office shack and commissary, presided over by one or more clerks, known to the men as inkslingers. The commissary carried work clothes, loggers' boots, pencils, paper, candy, etc., and most important of all, Copenhagen!

Snoose

"Snoose," sometimes called Nordic breakfast food, was essential in a sawmill or logging camp. Good food was important, but snoose was a necessity. Snoose was a combination of damp tobacco shreds, salt and flavor. To use it, the thumb and forefinger grasped a charge out of the box, conveyed it to the mouth, where it was deposited between the lower lip and the gum, just to the starboard side of the bicuspid tooth. Years of snoosing will burn out a cavity that holds the snoose in place. Snoose was never sniffed, as is done by people of eastern Europe, nor dipped, as it was in the southern states. It is potent stuff. The Scandinavians start using it at an early age and build up a tolerance that permits them to use large quantities without reaction. The result to the uninitiated is somewhere between being kicked by a mule and swallowing a half stick of dynamite.

Copenhagen came in little round tin boxes and the boxes were wrapped in paper to make a roll, ten boxes to a roll, as I remember. The amount used was tremendous. I read somewhere that one year, 40,655,395 pounds of the stuff was used in the United States. I don't vouch for that figure, but can believe it.

Some men nailed up the empty cans on a wall, and I have seen large walls of a mill, or the inside of a bunkhouse, entirely covered. Truly, the Scandinavians were the backbone of the timber industry, but they would have been worthless without snoose.

At that time, wrist watches were not in common use. Carrying a watch in a pocket with sawdust and other dirt was hard on a watch. An ordinary watch would fit neatly into a snoose box if there was a hole for the stem to stick through, so many men protected their watches by carrying them in a Copenhagen box while at work. A common question was, "What time is it by your snoose box?"

Near the cookhouse were the bunkhouses, close enough for the men to hear the clanging made by beating on an iron triangle, at getting up time or eating time. A machine shop adequate to keep the equipment in good repair was a must. In charge of the shop was the "donkey doctor," the man who repaired the donkeys (engines). The top man over all camp activities was the "bull of the woods."

One of the companies I had some contact with decided to open up a new tract of timber. This tract was on a railroad, far up in the mountains. Their engineering department was instructed to pick a campsite and draw up plans for a camp big enough to handle two hundred men. The site was picked and the plans drawn up to clear the site of trees and brush. Several carloads of lumber, nails and other building supplies were also sent up. When the site was cleared, and no plans on hand, the boss couldn't see letting that crew be idle, so he set about construction on his own. He built a cookhouse, dining room, office, storeroom, machine shop big enough to care for 20 steam donkeys, four locomotives and associated rolling stock, and the incidental structures needed.

About the time the job was finished, the plans showed up. As might be expected, the buildings did not exactly conform to the plans. The engineers wanted the construction changed to fit the plans, but the construction boss not only refused to make any

Wood-fired "steam pot" donkey. Later machines were much larger, used oil fuel. Cut-off "tin pants" and "corked" boots are evident in this Tillamook County crew. (OHS)

changes, he threatened to quit if anybody else did it. The company evidently thought a good construction man was worth keeping, for they made no changes. I guess the camp layout was functional, for it served until the tract was logged off.

Falling

The actual process of changing trees in the woods to logs in the storage pond is divided into two parts, cutting and logging. The cutting crew consists of fallers and buckers, both highly skilled workmen. Fallers work in pairs or "sets," and they do just what the name implies—they fall trees. Buckers work alone, cutting the trees into logs. The principal tools of the faller, in the early days, were ax and saw. A falling ax was a double-bitted ax, the blades longer and thinner than other axes. It weighed four and a half to five pounds, depending on the preference of the man. The handle was 48 inches long, sometimes with a little side bend so a man could chop a large undercut without skinning his knuckles. The handle might add a pound or two to the total weight of the tool. Keeping the edge of his ax was each man's responsibility. It was a necessary task and a matter of pride. The old story of an ax sharp enough to shave with is true—I've seen it done. However, a double-bitted falling ax isn't as convenient a shaving tool as a safety razor or an electric shaver!

A falling saw is a narrow, flexible band of steel, with teeth on one side and a handle on either end. The length varies from five to ten feet, depending on the size of the timber. Sharpening this tool was the responsibility of management. The saw filer was usually an older man with many years' experience as a faller or bucker, usually both. His skill was essential to keeping a good cutting crew, for usually they worked on a piece-work basis. The fallers' rate was based on the board feet of timber downed; the buckers' on the square inches of cut made. They called this "working by the bushel," or "bushelling." I understand that the power driven chain saw has just about eliminated the hand operated saw. This is a machine having teeth mounted on an endless chain, running over wheels on either end of a bar. The gasoline motor powering this is mounted on one end of the bar, the whole assembly light enough to be handled easily by a strong man.

Other tools of the faller are wedges, steel slabs about a foot

Photo above shows use of choker, cable placed around log to attach it to line. Choker is 1⅜ or 1½-inch rope core cable. In center is sliding hook (Peter hook), which accepted knob on end of cable (called a D), and would slide up to give firm grip on log. The D on the other end of choker hooked into socket on main line. (Courtesy R. Lee Johnson) Saw filer, below, works on crosscut saw in filing shack. (Courtesy J. T. Labbe)

long, three and a half inches wide and half an inch thick at the thick end. To drive these, he carries a sledge hammer, with eight or ten-pound head. Awkward to carry around, but necessary, were a pair of springboards. A springboard was a plank, about six inches wide and four feet long, with a steel angle iron bolted to one end. The steel knuckle could be jammed into a notch in a tree and the board used as a platform from which to saw or chop. Last, but not least, was the oil bottle, filled with kerosene, for cleaning off the pitch that would accumulate on the saw. This gummy material could make it impossible to drag a saw back and forth through a tree.

Falling was hard work. The men not only needed the strength to use the tools of their trade, but they needed to be agile enough to get out of danger fast on occasion. It was highly skilled work and there was much pride in accomplishment. The first step in falling a tree was to decide which way to drop it. Factors such as lean, wind, or an asymmetrical top, must be taken into consideration. Logs must lie either towards, or away from, the donkey, which was usually at the top or bottom of the slope. If a log laid crossways of the line of pull, it might hang up behind a stump. The next step was to decide where to cut. Sometimes large buttress roots made the lower part of a tree of little value for lumber, so the tree would be cut above them. This cut might be as much as 12 or 15 feet from the ground. To get the faller up to work that high was the function of the springboard. A notch would be chopped about waist high, and the knuckle of a springboard jammed into it. Standing on that, the faller would cut another notch and place a board on it. Standing on that, he would cut a third notch and bring the first springboard up to put in it. This process would go on until the workman reached the desired height. Both men of the team did this. Once the men were in place, the trunk of the tree was sawed, possibly a fourth of the way through, in the side toward which it was to fall. A large notch was chopped on the upper side of the cut, called an undercut. Then a saw cut was made from the opposite site to meet the first cut. If everything went well, shortly before the two cuts met, the tree would begin to lean. Then there would be the well-known cry of "Tim-m-m-ber." The men would scurry away from the stump, and with the sound of breaking limbs, another forest giant would come crashing down.

The shock of the fall would cause the very ground to tremble. Sometimes it was not that simple, and to get the tree to fall where desired, the wedges must be driven into the second saw cut behind the saw, to release a pinch on the saw, or to tip the tree in the desired direction. Sometimes a pebble would be placed in the undercut to change the fall of a tree. This was called "jumping" a tree. A good pair of fallers could place a tree anywhere wanted, no matter what the complicating factors. A show-off stunt was to set a peg in the ground where a tree was supposed to fall, then drive it into the ground by dropping the tree on it.

Buckers were usually older men and worked alone. Their job was to cut the tree into lengths that could be handled. The standard log was 40 feet, the length of a railway car. Defects such as knots, rot, breaks, etc., could sometimes be cut out to make a shorter, but higher grade, log. Their principal tool was a bucking saw, shorter than a falling saw, heavier and stiffer, because it had to be pushed as well as pulled. Other tools were the same as for fallers, except that the bucker did not have use for springboards. Most of his work was done close to the ground. That was the reason buckers were usually older men—they did not like to work on the boards. Deciding where to cut to get the maximum quality and quantity was a matter of experience and judgment.

Clothing worn by loggers was fairly well standardized. Essential were the boots—heavy, stiff, laced leather boots, with eight or ten inch tops. The soles were extra heavy and set in them were several sharp spikes or caulks (commonly called "corks"). These caulks enabled the men to keep their footing on slippery logs, but were murder on floors. The sign "No corks allowed," was often seen on the doors of business places around logging camps. The walks around camp, floors of commissary and mess hall were soon badly cut up.

Shirt and belt overalls were usually made of a heavy, water resistant khaki material. The trouser legs were cut off (stagged) just below the tops of the boots and worn outside the boot tops. The shirt tail was short, and worn outside the trousers. Because these garments were so stiff, they were commonly called "tin" clothes. The reason for the shirttail outside the trousers, and trousers outside the boots, was very simple. Much of the time, these men worked in the rain, and with shirttail and trouser legs

outside, water would run off like it does off the shingles on a roof. The hat was usually of material similar to the other clothes, badly battered from all sorts of unconventional uses. Underwear was of the "long handled" or "long John" type, frequently pure wool, because this was warm even when wet. Some men wore wool underwear the year round.

After the tree was made into logs, it still had to be transported to the mill. Getting a 40-foot log, containing 5,000 or more board feet, weighing possibly eight or ten tons, out of rough country was quite a problem. And this *was* rough country. The Coast Range is not high, but it has been badly eroded over the centuries. As one old logger put it. "When the good Lord made this country, He found he had too much material on hand, so, to get it all in, He had to stand part of it on end."

The key machine in getting out logs was the steam donkey. This was a steam engine of up to 200 horsepower, mounted on skids, called a sled, made of logs three to four feet in diameter. This engine was geared to a set of drums, usually three, on which were coiled wire cables. The larger cable, or main line, was from one and a quarter to two inches in diameter, depending on the job, and might be up to 3,000 feet long, though usually less than that. The smaller line, or haulback, would be one half to three quarters of an inch in diameter, and would be a little more than twice as long as the main line. It was run through a block fastened to a stump on the extreme edge of the tract, and its function was just what the name implies—it hauled the main line back out after a log had been pulled in. The smallest line was the straw line, or guinea line, three-eighths or seven-sixteenths of an inch in diameter. It was light enough to be dragged out to the haulback block by manpower, and was used to pull out the haulback. The reason for the name, I don't know. Nobody could ever give me a reason for it. Guesses were varied, but the most plausible one was that, compared to the main line, it was small as a straw. As for guinea, you guess.

The donkey, which weighed many tons, was moved around by skidding, dragging it on the log skids on which it was mounted. The end of the main line would be made fast to a stump or big rock, the line hauled in, and the machine pulled itself along. There would be a block on the front end of the sled, called a fair lead, and when moving, the main line would be threaded through it. When the main line was tightened, it tended to lift the front end of the sled so it would slide along the ground more easily. It was surprising where these massive engines would go; they would climb hills so steep that the puncher had to be tied to the machine to stay on it.

High Climber

If the logs were dragged directly to the donkey, this was called ground logging. This was possible in only a few locations; the stumps and rocks were obstructions to the passage of the logs. To get enough lift for the logs to clear these obstructions, the usual expedient was to use a spar tree. This was a tree left standing near the intended location for the donkey, called a landing. The preparation of a spar tree was quite dramatic and has been publicized in books and on the screen. It was done by a high climber, who worked his way up the tree. He climbed with the aid of climbing irons, similar to those used by telephone linemen, but with longer spurs, to reach deep into the bark of the tree. He also used a sling,

a three-strand hemp rope with a small wire cable through the middle. The sling was long enough to reach around the tree and the ends were fastened to a heavy leather belt.

The climber would go up as far as the sling would let him, get his spurs firmly fixed, lean forward toward the tree, give the sling an upward flip, lean back against the belt to hold the sling in place, climb up as far as the sling would allow, and repeat this routine until he reached the height decided upon, cutting off any limbs as he went. A few of the climbers used springboards in place of climbing irons and a sling, working up a tree by putting one board above the other. I'm not sure, but I think that was the way it was done before climbing irons were used.

Topping a spar tree was a dangerous maneuver, calling for a high degree of skill and strength. A man on a spar tree was entirely on his own, and had to meet, without help, any emergency that came up. If a sudden gust of wind caused the severed top to fall toward him, rather than the way he intended it to fall, he couldn't run. If the top of the spar split inside his sling when the top fell over, he would be squeezed. If his climbing iron spikes tore out, or if a piece of bark pulled off under his weight, his fall would probably be stopped by his sling catching on some rough spot, but he would get quite a jerk.

The accident that was most feared was cutting the sling. Part of the chopping had to be done within a few inches of this lifeline. The steel cable was incorporated in the line to make it more difficult to cut, as well as to stiffen it enough to facilitate the flipping maneuver, but even so, it was sometimes cut by a misdirected ax stroke. Seldom was the man able to catch anything to keep him from pitching backwards off the tree to almost certain death. If a man nicked his sling, he came down that tree really slowly and cautiously.

The climber carried with him, fastened to his belt, a short handled ax, a stiff, one-man saw, a light pulley called a passing block, and the end of a light line. When he reached the desired height, the passing block was fastened to the tree, the line threaded through it, and the helper below could send up anything more needed.

After the tree was topped it had to be rigged. This consisted of attaching the necessary guy lines, and hanging the main block,

Rigged and worked spar tree. Right, high climber starting up with tools hanging from belt. (OHS)

Top left, topping a tree; top right, "show-off" standing on treetop. Left below, well known high climber Pete Makovich (courtesy A. B. Berns); right, spruce so old it was dangerous to cut, so top was blown off with 80 sticks of dynamite wrapped in "bologna" around tree. At first detonated by long fuse, later detonator was put in snoose box and used as rifle target. Kinney Log Co., 1922 (courtesy R. Lee Johnson).

or "bull block." The size of this depended on the size of the main line used, and it might weigh several hundred pounds. The height of a spar tree varied with the terrain and the tree available.

I have heard of, though never saw, a spar 180 feet high. I think most were less than 100 feet. The cut-off place was not less than ten inches across, and such a tree would be at least five feet in diameter at the ground.

Sometimes it was decided that topping the desired tree in the usual way was too dangerous, maybe due to a lop-sided top, heavy wind, suspected heart rot, etc. Then a different plan was followed; the tree was cut off with dynamite. A piece of burlap, long enough to encircle the tree, was laid out on the ground and sticks of dynamite laid end to end along its length. Then this was formed into a roll, with each dynamite stick tied so it could not get out of place. This was called a bologna. The bologna was wrapped around the tree at the proper place, the fuse lighted, and everybody got clear, for there was no telling where the pieces might go. Dynamite left a ragged stump, so such a cut was made a few feet above the desired cut-off place. Then the climber could saw it off wherever he wanted.

A high climber had to be a resourceful man, quick-thinking and daring, as well as a man of considerable strength and woods experience. It also brought out any tendency to be a show-off. There are stories of climbers who would stand up on top of a spar and dance a jig, or clown a bit, even some that would stand on their heads. Such stunts were purely for show.

More functional was speed in getting down a tree. The man good at that could throw his hat in the air when he was ready to start down, and beat it to the ground. That was when loggers wore cloth hats; the hard hats worn by modern loggers might fall faster.

Climber Lars Larson one day was working on a tree with lots of limbs. He had cut off several, and they were lying where they fell about the base of the tree. When he was about 40 feet up, he made the capital mistake—he hit his sling with his ax and severed it cleanly. Of course he fell backwards, pivoting on his spurs, and plunged downward head first. Fellow workmen, seeing him fall to what they thought certain death, rushed towards the tree. Evidently the pile of limbs broke his fall, for before the others could get to the tree, he crawled out of the pile on his own power, shook his head a bit and said, "You can't kill a square-head." Then he fastened a spare sling to his belt, went up the tree, and finished the job.

One spar tree was near a road. While the high climber was at work, some people had stopped and were watching. It was not often such a job was done where non-loggers could see it. The climber's helper, a loud talking, officious individual, was continually saying, "Stand back, stand back, he might drop something. Stand back, you might get hurt." While he was doing this, a large chip from the undercut came flying through the air, struck the man on the arm and broke it. The bystanders were quite willing to believe he knew what he was talking about, but most of them thought it was hardly necessary to go to such lengths to prove that the area below a high climber was dangerous.

The spar was well braced with cables in all directions, one set fastened near the top, to keep it from being pulled over, one or more sets near the middle to keep it from buckling. A heavy pulley, or block, was hung at the top of the spar and the main line threaded through it. If the block was big enough to pass the choker hook, it was called a Tommy Moore. Never found anybody that knew why.

If no suitable tree grew where a spar was needed, one would be brought from someplace else, raised, and anchored in place.

Ted Berns, one of the old time logging bosses, told me this one.

"We had a tree to rig in a tough place and it had to have an unusual arrangement of rigging. My head rigger was a new man, and I didn't have much confidence in him, so explained very carefully how I wanted it done. He listened until I was through, then said, 'I quit.' I was surprised and asked why.

" 'I laid awake half the night last night figuring out how to rig that tree, and now you come along and tell me some other way to do it. I just won't do it.'

"I thought if a man was enough interested in his job to spend half the night figuring how to do it, he should be encouraged, so I told him, 'Go ahead. Do it your way,' and he did a good job. Worked fine. I kept that man until we finished that show and he was a good man."

A short length of cable called a choker was looped about a log and hooked to the main line to pull it in. Sometimes it was necessary to dig out a tunnel under the log to get the choker around it in the proper place, and the easiest way to dig it was to shoot it out with dynamite. One man, called the powder monkey, carried several sticks of dynamite and a roll of fuse in a gunny sack and detonators, "caps," in his pocket. He would shove an iron rod down to what he thought was beneath the center of the log, wiggle it around until the hole was big enough to take a stick of dynamite. If the charge was properly placed under the log, it would blow out on both sides, leaving a hole big enough for a man to crawl through, pulling one end of the choker with him. Some of these powder monkeys were unbelievably careless with the dynamite, or more particularly with the caps, and sometimes it caught up with them. When it did, I had either a big job or no job at all. They never had a small accident.

One choker setter was in a tunnel when he felt the log settle. He immediately began to yell for help. Other men came up; they got jacks to lift the log, and started the laborious task of jacking up several tons of log. Then one man noticed that the log was not touching the man in the tunnel. He had felt the log settle, assumed he was pinned, and had made no effort to crawl out. The other fellows were not happy to have carried those heavy jacks all the way from the donkey to free him from a hole where he wasn't caught!

Some "high-ball" camps put a lot of pressure on the men to hurry. Sometimes the men claimed they were pushed so hard the chokers got too hot to handle with bare hands.

Donkey Crew

To operate a donkey, two men were needed, the donkey puncher or engineer, and a woodbuck. The puncher was the key man of the operation. If he could answer signals promptly, knew how to ease logs through the tough spots, and speed up over the good ground, he could make the side show a profit. It was essential that he always have a mental picture of what was going on out in the brush, so he could be of maximum assistance to the rigging crew. Several years' experience on a rigging crew was almost a necessity.

The woodbuck's job was to cut the wood and feed it into the fire pot fast enough to keep up steam. Whenever it was needed,

Typical "steam pot," wood-burning donkey. Note hand-hewn sled timbers, three drums, and "tail hold" to background tree. Lee Tittle is left, holding block; his brother Oscar third from right (Tillamook, about 1912). Courtesy Tillamook Pioneer Museum.

a log would be dragged up and dropped, handy for the woodbuck. If it was possible to find a dead or broken log to cut up for wood, it was pulled in and used. If not, a regular production log was used. These logs were a frequent source of argument. The woodbuck wanted a nice, straight grained log, for it was easy to split. Also, he could not get the big knots into the fire box, so they had to be carried out of the way. Management wanted the good logs to go to the mill, and the poor grade logs cut up for wood. They never could get together on that point.

The conversion of wood-burning donkeys, or "steam pots," to oil burners, and later, the use of gas or diesel internal combustion engines, spelled the doom of the woodbuck job, just as technical improvements have doomed so many other jobs. Then the caterpillar tractor largely nosed out donkey logging.

The crew operating with a donkey was called a side and was bossed by a "hooktender." Under him was a "head rigger," whose responsibility was to see that the equipment was properly set up and kept in operating condition. There were two to four choker setters, men who placed the chokers and hooked them to the main line; a powder monkey; two chasers, who followed the logs in and changed the position of the chokers if necessary to get the log around obstructions; and a whistle punk.

As it was usually impossible for the crew to see one another, all the donkey work was done on signals. These were given by the whistle on the donkey, actuated by a wire stretched to a workman, called a whistle punk, perched on a high stump, or other elevation, where he had a view of the crew. Men on the ground signalled to him with arm signals, similar to those used by a football official, or a railroad brakeman, and he translated the arm signals to a series of long or short blasts of the whistle. Everybody could hear these, knew what they meant, and could take proper measures for safety. The signals were standardized for the industry, so men moving from one logging show to another were never in doubt as to the meaning of them.

A heartstopping run of six long blasts of the whistle meant "Somebody's hurt," and was the signal for stopping all activities until the injured man was gotten out. If it was a fatal accident, work was over for the day. Wives and mothers of workers living close enough to hear the whistles would often gather around the camp office until it was known who was the victim. They were a nuisance, but one couldn't blame them.

Although men on a side were hired for specific jobs, they were expected to do anything that needed to be done, and most of them could. They were good all-around workmen and were proud to be called loggers. Rigger, puncher, setter were secondary terms. I understand specialization has now come to the woods, as to most other industries and the good all-around logger is rapidly becoming an extinct species of animal.

Recently, while talking to Ted Berns, he told me that not long before, a hooktender who got his training in old time logging, called him and said, "Ted, do you know where I can get some good loggers?"

"I have been out of the game several years," Ted replied. "You ought to know more about it than I do. Don't you know where there are some good men?"

"Sure I do, but I'd have to use a shovel to get one."

As these men worked hard under rugged conditions, their humor was also rough. A favorite prank was to take a Kotex pad, dope it up with Mercurochrome from the first aid kit, and slip it in a man's lunch box.

Dan Murphy was a superstitious chap, not too bright, though a good, dependable woodbuck. One nice Sunday afternoon, he wandered off by himself, found a quiet, sunny place, and stretched out for a nap. Some other men found him, got some gasoline, poured it on the ground in a ring around him, and struck a match to it. When the gasoline was burning good, they waked Dan by throwing a rock at him. He roused up enough to see the flames all about him, and started screaming, "I've gone to hell, I've gone to hell, I've gone to hell."

When the flames died down and he saw the laughing bystanders, somebody would have gotten a licking if they had not held him.

Logging Railroads

After the logs were brought up to the railroad by donkeys, they still had to be transported to where they could be dumped into the water. For some reason I do not understand, my memory of logging locomotives is almost a blank. I know that they were of a special type, designed to operate on steep grades. All those I can remember had vertical pistons. This helped give the engine an unusual appearance. What their other peculiarities were I do not remember. Maybe I just was not interested in locomotives.

Railroads were in common use in logging when I came into the picture. I have been told that Bill Berns, Ted's older brother, brought the first locomotive into the country. He bought it in Portland, moved it by scow down the Columbia River and around to Nehalem Bay. He unloaded it from the scow at Wheeler and moved it to God's Valley, ten miles or more. He laid rails ahead of the engine, then picked them up after it had passed over them, and carried them around in front again.

The rolling stock used was of two types, standard log cars, used if part of the haul was over a regular railroad; bunks if the logs were hauled all the way on private road.

Log cars resembled ordinary flatcars, except that there was no deck, the trucks being connected by steel beam. At each end,

Incline above built by Whitney Co. while logging on Kilchis River. Cars were moved up and down by large donkey engine at top. (Courtesy G. B. Abdill) Below, Coats Co. crib bridge southwest of Tillamook, 350 feet long, 110 high and 40 wide at bottom. So far as known, largest crib bridge ever built; 750,000 feet of lumber was salvaged from it when dismantled. Donkey on car is hauled by Climax locomotives. (Courtesy R. Lee Johnson)

over the trucks, was a heavy steel bolster to carry the logs. Also, logs were frequently hauled in ordinary gondolas.

Where all the haul was over private line, the customary equipment was the log bunk. This consisted of four wheels like those on one end of an ordinary car, with a bolster over them. These bunks were spaced far enough apart to accommodate the logs intended for the load. Sometimes the load would be just one big log, sometimes two smaller logs on the bunks, with a third, a "peaker," on top. The weight of the logs would keep the trucks from pulling apart. That seemed peculiar to me but I understand that such an accident rarely happened. Using these bunks allowed logs of any length to be brought out, and I understand it was easier to maneuver the empties. I have been told that one company on the Columbia River specialized in long logs, used for keels and masts in the old wooden vessels. For such extra long logs the bunks were essential. The Tillamook roads were all too crooked, because of the rough terrain, to permit extra long logs being brought out.

In the twenties, the Whitney Company had a railroad up the Kilchis River. They dumped their logs at Idaville, a couple of miles south of Bay City. The road went up the north side of the valley to tap a tract of timber they owned. They had two camps, White Star and Blue Star. More than half of this timber was on a high plateau, and to get to it, they built an inclined railway, 4,800 feet long, at an incline of eighty per cent. Cars were pulled up and let down this incline by an endless cable, powered by a donkey at the top. White Star camp was at the bottom of the incline, Blue Star at the top. The latter was headquarters for the operation.

They had a foreman up there who stuttered. He was a good logger and handled men well. As is usual, the harder he tried to talk, the worse he stuttered. One day when the superintendent asked a question, he tried to answer. The super was a quick talking, impatient man, and was obviously annoyed that he could not get an immediate answer to a simple question. At last Buck said, "D-d-d-damit, if you would j-j-just l-l-look the other w-w-way, I'd t-t-tell you."

They had quite a fire near White Star and as usual all the surrounding camps sent men to fight it, and all possible men in the community were brought in. This crew was put in charge of a well known logging boss. Arrangements were made to bunk and feed the men at Blue Star, because White Star could not handle so large a crew, and besides, the fire might get to it.

The crew fought fire as long as they could see, then assembled at White Star. The man in charge called the superintendent and asked that cars be sent down the incline to haul the men up. The answer was "Oh no, we couldn't do that. We never let anybody ride on that incline, it's too dangerous."

"Listen, Tom, these men have fought fire all day, and they are tired. It's a damn sight easier to walk out to the highway than it is to climb that hill. If you want this crew here in the morning, you better send down them cars."

The cars came down.

This operation was later wiped out by a big fire.

The power involved in hauling in logs was tremendous. If the far end of a mainline were pulled to one side to be hooked to a log, when strain was applied, the whole length of the line would snap straight with terrific force, flattening any vegetation (or man) in the way. The necessity for keeping far enough from a line so that no possible side snap could reach him, was so deeply ingrained in every logger's thinking that it was commonly said he was "caught in the bight of a line," or "caught in a bight," to describe almost any kind of unexpected trouble. I learned early the force involved. A call came to pick up an injured man in a camp. A main line had broken and the free end whipped around. The injured man had been standing near a small tree, 12 or 14 inches in diameter. This tree was snipped off as if by a pair of giant scissors. The free end of the cable had just touched the man, nipping a piece of muscle, about two inches deep, from his thigh. If he had been a few inches closer, it would have taken his leg as easily as it snipped off the tree.

The power of the donkeys could move these large logs a half mile or so in just a few minutes. It was a spectacular sight to see a huge log sometimes striking a rock or stump and flying through the air, again almost burying itself in the dirt, but relentlessly crashing along through the brush. Sometimes two or more donkeys might be strung out in a line, one delivering logs to the next, the last one piling them up, or "cold decking" by a railroad loading switch. From this cold deck they were picked up by giant tongs, similar to ice tongs, but about six feet high, with points about three inches broad, and swung aboard a car.

One day I saw a fellow who somehow got himself caught by the tongs. One blade cut through the skin over the pubic bone, the other just below the belly button, the points meeting in the muscles of the abdomen. Through a confusion of signals by excited fellow workmen, he was swung through the air by this grab for a few minutes. When I saw him, the strip of skin and other tissue between the cuts was black and lifeless, so all I could see to do was cut it off. At that time, plastic surgery to repair a tissue defect was done by only a few men in the large city hospitals—we never thought about doing it.

By bending his body, the edges of the wound could be brought together and sutured. The wound healed nicely, but it was many weeks before he could stand up straight. After he finally got enough belly wall to straighten up, he was as good as new, except that he had a very peculiar pattern of pubic hair.

Mention has already been made of the important role of the railroad in this phase of logging history. It was necessary to get the rails to where the timber was. The ingenuity and engineering skill used in building and operating these roads in that rough country was amazing. Special techniques of trestling and bridging were developed. Locomotives designed for that type of road were built, totally different from "over the road" locomotives.

As trucks became more efficient and dependable, they outdated the railroad. Now, I understand, in some places helicopters are taking over the log transport tasks. Progress? Maybe.

More About Loggers

Logging accidents were fairly common, and their type reflected the occupation. The cutting crew, working with edged tools in standing timber, was more likely to get lacerations or to be hit by a falling dead limb, called a "widow maker." The logging crew, working with heavy equipment and handling heavy material, was more likely to have crushing injuries or fractures. Fatalities were not uncommon.

One of the disagreeable tasks that frequently fell to my lot was to break bad news. Everyone seemed to think it was part of my job. The common excuse was "I'm just a dumb logger; you can talk better than I can. You gotta tell her." How do you tell a woman tactfully that her husband, or her son, whom she sent off with a full lunch pail a few hours before, has been killed in an accident, especially when the family are neighbors and friends? How do you approach the job? What do you say? I had to do it a lot of times; it was just as hard the last time as the first, and I never worked out a satisfactory technique for the task. I hope I never have to do it again, but if I do, it will be just as bad an experience as it was 40 years ago.

In the thirties, the State of Oregon waked up to the accident rate in the woods, and decided to do something about it. A safety code was set up and a group of hard-nosed inspectors was put in the field. I do not know about other areas, but in ours the inspector was an old time logger, who had been cracked up in an accident, so that he was no longer able to work in the woods. He knew logging and he knew loggers. Also, he knew from experience the result of accidents. I knew of one instance where he caught his best friend violating a regulation through misunderstanding its interpretation. He told him: "I think you made a mistake, so I'll forget it this time, but if I ever catch you doing it again, I'll put you in jail." He really meant business. Do not know about the rest of the state, but in our area the accident rate dropped off sharply.

Loggers were a superstitious group. One summer, a crew working in a deep narrow valley, had two or three fatal accidents, rather close together. Someone reported that an eagle had flown over before each accident, so they concluded the eagle was a warning. Thereafter, whenever the eagle flew over, the crew laid down tools and signalled for the "mulligan" (man train), to pick them up. Production was off that summer. I suspect the eagle had a nest around there somewhere, and when they saw her she was looking for something to feed a nestful of hungry babies—maybe going to the coast for a fish.

During our time, the famous Tillamook Burn of 1933 occurred. The heat generated by pulling a log across a dead windfall cedar caused a blaze to start, and though there was a crew with tools on the spot, weather conditions were such that they were unable to control it. The fire spread until it had covered 380 square miles of some of the finest virgin timber in the United States, destroying about ten billion feet of lumber. The smoke was so dense that 20 miles away, at noon time, it was as dark as evening; cars had headlights on full, chickens went to roost. The heat was so great that large pieces of burning material were caught in the updraft and carried for miles to set other fires. In many places, the topsoil actually burned down to the bare rock, and the exposed rock surfaces were fused. Iron in buildings or equipment in the path of the fire melted down to shapeless lumps. The scars of this fire can still be seen from the road, more than 36 years later, though new vegetation is beginning to take over.

As a result of this experience, laws were passed sharply restricting logging in dry weather. Each camp was required to install a hygrometer, and when humidity dropped below a certain point, stop work. The loggers bitterly resented this; they could not understand how any gadget was smart enough to tell them when it was not safe to work. The first one I saw I did not recognize and asked what it was. "That is the damned machine that makes the humidity," was the reply.

I helped measure an old cedar snag that was well known to the loggers and surveyors of Tillamook County. It stood almost on a section corner, a short distance east of the north end of Lake Lytle, and was a well known "witness tree." The top was broken off, possibly 40 or 50 feet from the ground, and was not present.

It had probably been worked up into shingle bolts sometime in the past, so we had no way of knowing how tall the tree had been. It was hollow also, and the winter before, the inside caught fire and burned out quite a hole. This fire occurred in the middle of a heavy rainstorm. It might have been caused by lightning, though nobody had noticed any flashes, and lightning is very rare during the winter storms in Tillamook. A pair of surveyors had been working in the area a week or ten days before, and had built a little fire to hot up their lunch coffee. They said they built this fire 75 or 100 feet from the tree, and it had burned out before they finished eating. We decided their fire had been over a root of the cedar. An underground root will sometimes smoulder for several days, and in this case, we think that is what happened; the root went deep enough that the fire was not put out until it had eaten its way to the heart of the tree. There it found a little crack in the dead heart wood that acted as a flue, and it burst into flame, burning until the rain storm put it out.

There were large buttress roots on all sides of this tree, and we tried to get above them by holding our measuring tape as high as we could, possibly seven feet above the ground. At this place, the circumference was just over 72 feet.

This snag has since been cut down and sawed into shingles, but the stump is still there and I suppose it will be there 200 years from now, if nobody takes the trouble to burn it, or blow it out.

A logger was working in the White Star camp. He was a young fellow with more than average energy and ambition. One day he showed up in the yard outside the office and called me to the window. He was redolent of "eau-de-skunk." His story was about as follows. He had been working at White Star for several months. He noticed there were quite a number of fur bearing animals in the area, so got a bunch of traps and put them out. He would get up before daylight to run his trapline before breakfast. He was doing pretty well, too, making almost as much from fur as from his job.

The morning before this conversation, he was returning to camp along a narrow deer trail through an old burn which had grown up with very dense brush. In fact, the brush was so thick it was almost a solid wall. By the dim early light, he saw a large skunk approaching him. No person or animal will dispute the right-of-way with a skunk, so Clarence pushed off the trail into the bushes enough to let Mr. Skunk get by. It was a beautiful pelt, one of the finest he had ever seen, and to have it walk past, touching his feet, was more than he could stand. It suddenly came to him that if he grabbed that animal by the tail, held it at arm's length ahead of him, and ran so fast the skunk could not get turned around to train its artillery on him, when he came to a more open place he could give a big swing and bang its head against a tree. The little point he overlooked was that it was so far to an open place that he could not run that far fast enough to keep the skunk's weapon turned away. When he had to slow down for want of breath, the skunk managed to get turned around to the proper position and gave him both barrels at close range. By the time Clarence could get his eyes open and begin to breathe again, of course the skunk was gone, but there remained conclusive evidence it had been there.

He used his hunting knife to dig a little hole in the ground, in the edge of the clearing around camp, buried his clothes, then sneaked into the bathhouse and took a long, hot shower. When he tried to get into the bunkhouse to get more clothes, the boys threw him out, then threw out to him a set of clothing. When he went in to breakfast, the cook chased him out of the mess shack, telling him if he would get clear away for fifteen minutes, he would find his breakfast on a stump at the edge of camp. He wanted to go to work, but the boys wouldn't let him on the mulligan; besides, his partner said he wouldn't work with him if he walked up to work. By that time he had concluded they didn't want him around, so he decided to come to town. The only way the train crew would let him aboard was his promise to stay on the last car. When he got down to the highway, nobody would give him a ride. As he walked down the road, everybody got away from him as quickly as possible.

What he wanted of me was something to deodorize him. The only thing that came to my mind was to suggest he try Clorox. (He told me later it didn't help much.) He had a married sister in town and she agreed to let him use her bathroom for a series of Clorox baths, if he would climb in and out of the window, leaving the window open after he left. She locked the bathroom door on the outside, and her family attended to their bathroom chores at a neighbor's for a couple of days. He went back to work in a few days, but said running that trap line was no fun anymore. However, he felt he had to keep on with it to make up for the time and clothes that skunk had cost him. A point he apparently failed to appreciate was that the skunk was acting in self defense. A skunk never starts a fight, his armament is purely defensive. He tends to his own business; he doesn't go out of the way to make or buy friends; he has no enemies; nobody tries to eat him and everyone respects him because he stands up for his rights. There ought to be a good editorial there, and maybe a lesson in international relations!

One day while talking with a logger, a fellow workman was mentioned. This man had recently lost a little finger in a careless moment with a falling ax. The logger commented, "Of course it's a shame to lose a finger, but he wasn't too unlucky. All you can do with a little finger is scratch your ear, and he can do that with a match stick."

One camp I served was about 30 miles from Garibaldi. More than half this distance was over a state road, which by this time was pretty good. The rest of it was mountain road, much of it single track. If two cars met, one had to back up to a turnout before they could pass. I found out that the boys in this camp had been getting up a pool when I was called, each one throwing in two bits and each registering a guess as to how long it would be from the time I was called until I got there. By the time I got on to this little game, Jimmie, the camp timekeeper was winning pretty regularly, sometimes, just after pay day, rather sizable pots. After the camp closed, he told me why.

He figured anybody could guess pretty close to how long it would take to drive that distance, taking into consideration weather and road conditions, but there were two uncertainties: 1) how long it might take me to get started after getting the call; 2) the number of cars I might meet on that single track road, and how much time would be lost in backing up. So he set about removing the uncertainties. 1) They could hear the accident whistle (six long blasts) a long time before they knew the severity; whether I should be called or the man could be brought to me. If the accident was severe, a man would be sent ahead of the stretcher to tell the donkey puncher at the loading siding to telephone down to the office

Logging companies set a good table. (J. F. Ford photo, OHS)

to call me. That would give me time to get there by the time the injured man arrived at camp. Sometimes it was a matter of 30 minutes or more from the time the accident whistle was blown until this messenger could reach the railroad phone. So, when Jimmie heard the whistle, he would slip out of the office to his own private phone, invest 20 cents of his own money in a long distance call, tell me the accident whistle had blown and he would call later with details. That gave me a chance to take care of anything pressing, and be free to jump in the car and take off if necessary. Nobody at camp knew of the first call; they only heard the second, which was made from the office in the hearing of anyone around, after the severity of the accident was known. This call started the clock on the pot. 2) His solution to the second problem was to stop all cars going the wrong way with, "We've had an accident and the doc's acomin', an' he will probably be drivin' like Hell. Better just pull over and wait until he gets here." It usually worked.

I thought he deserved those pots!

Food

Food served at the camp cookhouses was always good and plentiful, but at one camp it was especially superior. It was so good that I showed up there at meal time whenever I could possibly find an excuse to do so. One morning, all the Finns in the crew didn't get on the mulligan, and shortly after showed up at the office with their turkeys (bedrolls) over their shoulders, ready to walk down the road. They demanded their time. After each man had been paid off, the clerk asked why the exodus.

"The grub been't fit to eat."

That was a surprise, so he pursued the subject further and found that they were sore because there was no rye flour hardtack. When he was sure that was the only gripe, he asked if they would go back to work if assured an ample supply of hardtack. That was agreeable, so all were rehired and a man was sent to Tillamook to buy up all the hardtack in the stores, enough to last until a stock could be obtained from the wholesale grocery house. Thereafter there was always a big stack of the large cartwheels of hardtack at every meal.

Food was served on long tables, with a backless bench along each side. The call to meals was beat out on a large triangle hung outside the kitchen door. The men came in promptly, sat down wherever they pleased (though most had favorite places), and started eating without ceremony. Every man devoted his entire attention to the task at hand, eating steadily, often noisily, never speaking except to ask for something he couldn't reach, and paying no attention to his neighbors, except to pass them food. As soon as he finished, each man got up and left.

The table would be covered with huge platters of food. There were usually three or four kinds of meat, several vegetables, a number of kinds of bread, pies, cake, cookies, etc. As fast as the dishes were emptied, the waitresses brought refills. The quantities some of these men could eat were astounding. It was nothing unusual to see a man pile on all the corned beef and cabbage his plate would hold, and then take seconds and thirds. One man I knew would eat a dozen eggs at a meal. If there were not at least six on the platter when passed to him, he would not take any. "Not worthwhile startin' on less'n' six eggs," was his position. He would wait for a fresh platter, and pass the time eating whatever else was within reach. Coffee, tea and milk (when available), were served in large, heavy mugs. Most men took several spoonfuls of sugar per mug of coffee or tea. Cream was evaporated milk, served in the can, with two holes punched in the top. Fresh cream was not wanted.

One peculiar custom that I think was general in mill and camp dining rooms, was the rule of silence while eating. I was told that the reason for it was that the waitresses were usually young women, and logger conversation was sometimes rather gamey. To avoid embarrassment, the rule of silence was imposed. Some of the girls were able to take care of themselves. As one was leaning over to place a large bowl of gravy on the table, the chap alongside reached down and pinched her leg. Without hesitation, she brought the large bowl of gravy down on his head so hard it broke and hot gravy poured all over him. It must have taken quite a blow to break the heavy stoneware bowl! Anyone but a hard headed Swede would have been knocked out. He was known as Gravy as long as I knew him.

One prank tried occasionally, but seldom twice by the same man, was to bring a harness buckle to the table, pretend to find it in the roast beef, and then loudly proclaim that the cook was serving horse meat, and hadn't taken time to remove the harness. Such a trick would rouse the ire of the cook, and the wrath of a cook could be formidable.

Breakfast and supper were served in the mess hall, but most men carried their lunch. On a side table in the mess hall, were the "makings," and as soon as he finished breakfast, each man "built" his own lunch. On this table were sacks of sandwiches, three to a sack, two meat sandwiches and one of something else, such as jelly, peanut butter, cheese, etc. A man could take as many sacks as he pleased and some took two or three. There was also on the table pickles, fruit, sausage of various kinds, cheese, dried or salt fish, pies, cake, cookies, etc., and each man was privileged to take any or all of them, as much as he wanted. Some lunches looked like enough for a group picnic. There was also coffee and tea. A few Thermos bottles were beginning to show up, but most men used metal canteens or large glass bottles. Whisky bottles were easy to come by. Most of the men had a tin can cached away where they worked, and come noon each would make a little fire and hot up his coffee or tea. If he couldn't find a safe place for a fire, or if fire warnings were posted, and there was any chance of a fire warden coming by he drank it cold.

The charge to the men for this sort of eating was usually one dollar a day. Feeding a logging crew was a non-profit business, even at prices prevailing then, but it was impossible to keep a good crew without good grub.

Before the time when refrigeration was general, food poisoning was a camp scourge. When it hit, it usually hit hard. Camp sanitary facilities broke down completely.

Sleeping accommodations were provided in bunkhouses, built to accommodate six to twenty men. Double decker bunks were built around the walls. In the center was a big stove, and around it there were a few tables big enough for a card game. Some of the better camps had cupboards or wooden lockers for clothes, but most of them just had nails in the walls. The bull cook kept a supply of firewood piled by the door, and swamped out the floor occasionally. Each man supplied his own bedding, sometimes blankets, but usually quilts, known as soogans. Some camps furnished straw for the bunks. There was also a bathhouse, where there was an ample supply of hot water for showers. Keeping it hot was another duty of the bull cook. For the use of the bunkhouse and bath, the usual charge was fifty cents a day.

Most of the loggers were fairly particular about personal hygiene, and the shower room was well patronized. But there were some not so particular. Ed Bywater was one of the latter group. He insisted that baths were weakening and should never be taken. He practiced what he preached, and the result was he always smelled a bit high, sometimes more than a bit. Once the others in the bunkhouse could not take it any more. They caught him, stripped him, carried him to the shower room, stretched him out on the floor, turned on the cold water (it was January), and used the bull cook's broom and scouring powder to scrub him. They did a good job. When he reported to my office to get various abrasions fixed up, he had only a very faint aroma. The story he had cooked up to explain the abrasions was a masterpiece, but as of now I can't remember it. Needless to say, the true story went around camp in a hurry.

Hiring Hall

There was a large turnover in employment in the camps. It was a joke, which was too near the truth to be funny, that it took three crews to run a logging camp—one coming, one working, and one going. Of course some local men were employed and these tended to be steady workers. The supervisory crew that lived in camp and had their families with them were permanent. Most of the men were hired through hiring halls in the nearest large town, in our case, Portland. The men paid a fee to be sent on a job. It was commonly rumored that camp foremen got a kickback on this fee, though I never saw any evidence of it. However, the shortness of the average employment looked suspicious. As the unions got stronger, the turnover in camps became less. Since most of the hiring halls in Portland were on Burnside Street, a common form of discharge was, "Bud, I'm transferring you to the other side." Greenies thought it meant transfer to another crew, but soon found out that the side to which he was transferred was Burnside.

The hiring halls located in Vancouver, Seattle, Portland and San Francisco were interesting institutions. Some of them were clip joints, where the object was to get all the money a man had, then lay him cold with knockout drops and sell him to a shanghier to deliver to a shorthanded ship ready to sail. More of them were honestly run outfits, to which the men returned time after time. In front was a large board on which were listed jobs available. Inside might be more notices of less desirable jobs. Most of the place would be filled with chairs and tables for cards. There was frequently a bar. Men looking for work would hang around, drink beer, play cards, visit with friends, etc., watching for a notice of the sort of job desired. There were not only loggers, but farm hands, sawmill men, railroad men, sailors, tie cutters, construction men, etc., as well as common laborers.

One accepted pastime while waiting was spitting at a crack for nickles. A good spitter could pick up enough change to pay his bar bill. Naturally, the conversation around such a place centered on their work, and probably more and better logging was done in the hiring halls than in the woods.

In the back of the building was a large chicken wire locked cage where a man might store any belongings he did not want to take to his room in a nearby rooming house. Since these houses had a bad reputation for thievery, almost everything except the clothes he had on went into this lockup. Such storage was not only safe, it was security for a loan if necessary. A man could have mail sent to this hall, meet friends, wait for a job to turn up, etc. It was a sort of poor man's club. When a suitable job came up, the man was shipped to it, the employer paying the fare if the man had a bedroll. No bedroll, no job.

When Edward came to the throne of England, a funny joke to ask was: "Does a man have to have a bedroll to get a job of 'kinging'?"

In return for all this service, the operators of these hiring halls, or "job sharks," got a percentage of the first month's pay check.

Employers would usually provide transportation to the job, but when they quit, or got fired, the men had to get back to the hiring hall the best way they could, often on shanks' mares, carrying all their possessions rolled up in their soogans, slung over their shoulders and across their backs. The roll, or pack, was known as a turkey, or a bindle, and the man carrying it was a bindle stiff, or a hobo. There was a sharp distinction between a bindle stiff or hobo, and a tramp or bum. The former was really looking for work and would work hard if given a chance, while the latter would not take a job if he could beg or steal enough to eat. If a tramp was hungry enough to take a job, he would do just as little as he could get by with, and probably steal anything that was not nailed down.

Sometimes the bindle was a long roll, carried over one shoulder, the ends tied together under the opposite arm. Sometimes it was a rectangular bundle, carried like a knapsack. I never saw anyone use a pack-board. The fact that all their personal property ultimately had to be carried discouraged accumulating things.

A certain class of worker would ship out on a mill job, stay a few days, and quit; the reason, any, or all, of the following, or something else. Weather too hot; weather too cold; weather too rainy; lumber too heavy; pay too light; boarding house grub too greasy; boarding house coffee too weak; boss too tough; too many Swedes; too many Finns; too many Indians; work day too long; not enough work; prices in commissary too high; quality of commissary goods too low; too many of boss' relatives in crew; too far from Portland; too close to Portland; bunkhouse too hot; bunkhouse too cold; timekeeper crooked; timekeeper incompetent; boss has pets; beer in tavern flat; too far to walk from boarding house to job; etc., etc. So he would go back to the job shark for another job. He was always looking for the perfect job, with perfect pay, perfect weather, perfect working conditions. Some foremen claimed they could spot them as soon as they walked on the job. They were commonly known as "sawmill inspectors."

Gyppo

Ole Hanson was a big lanky Swede, a powerful man, with very little education. By sheer brute strength and drive, with a lot of native shrewdness, he had managed to put together a little "gyppo" logging outfit, employing maybe a dozen men. Ole probably did more work than any two of his employees. A gyppo logger was a little operator who logged for a bigger outfit on contract basis. Seldom did they ever get much money ahead. They operated on a shoestring, and if anything went seriously wrong, they were not able to meet their obligations. Therefore, labor and suppliers got gypped. The companies employing gyppos frequently did the bookkeeping for them, for seldom did the gyppo have enough education to keep satisfactory records. I happened to be in the company office one day when Ole walked in. The head bookkeeper jumped him. "Ole, I haven't had a payroll or a scale sheet from you in three weeks. How do you think I can keep books for you if you don't turn in the information?"

Ole looked surprised and said mildly, "It bane good vetter."

"What in —— has the weather got to do with your reports?"

That was such an astonishing question that Ole exploded. "Hell, mister," he yelled, "when she is good vetter I vork; I keep my books ven she rains." And he stalked out.

There were two independent loggers operating above Foss, on the railroad. They were close together, logging in the same block of timber, and there were obvious reasons why it would be well for them to combine. They dickered for a time, then the deal fell through. When one of them was asked why no deal, he replied, "The difference between us is that I'll admit I'm crooked, but George won't."

Modern methods and equipment have revolutionized logging. The power driven chain saw enables one man to saw as much in a few minutes as two men could do in hours. The internal combustion engine, installed in trucks, donkeys, and bulldozers, has eliminated the railroad and the steam pot. By the same road that trucks haul logs out of the woods, workmen can drive in, so the men can live at home and drive to work, eliminating the camps. This has made a change in the type of men.

More loggers are married and live with their families. Many own their homes, with small acreages which they work on days off. They participate in community affairs. The men hold their jobs longer and are more dependable. Most of the kids have never seen a bindle stiff and would not recognize one if they did see him. They probably could not define the term "bindle stiff," nor differentiate between a hobo and a bum. Increased mechanization calls for different skills. As an oldtimer put it, "There ain't no more loggers—they're mechanics wearing loggers' clothes. And it all started when loggers started sleepin' between sheets."

Another said, "Loggers is gettin' civilized."

Cruising

Louie Terwilliger was a timber cruiser. One day he told me this tale:

"When I first started cruising, I worked for Sam Elkins. Sam was an old time cruiser and a good one. He had cruised for several of the big outfits for a long time, but in his later years was crippled up with rheumatism, so that he only did the easy ones himself, where he could get in easy and where the ground wasn't too rough. On the tougher jobs he sent me.

"He got a job to cruise a tract in the Siletz country, pretty high up in the mountains. It was getting along late in the fall, but he figured we might get it done before winter set in if we were lucky. Don Williams, my compass man, and me went to Toledo on the train, and shipped in our camp equipment. We bought supplies for a month and started looking for a way to get them out. The map showed a country road to about 14 miles of the tract. We soon found out that it was a fair road for a few miles, but the last 10-12 miles were just a track chopped out through the woods, up some hills so bad a team could hardly get over them in good weather, impossible when it was muddy. The only people on it were three or four stump ranchers. At that time of year, nobody was very anxious to tackle that road, afraid a rain might catch them up in the hills, but we finally found a young fellow with a team and buckboard who was willing to try. He got us, our equipment and supplies, up to where we were to strike out across country. One of the stump ranchers on the road had some horses and he jumped at the chance to make a little cash money packing our stuff in. It was pretty hard going, up hill all the way, some of it pretty steep. The first part of the way we followed an old skid road, and evidently the deer had used it too, for there was a trail of sorts. It was a cold, gloomy day, looked like snow. Time we got to where we were going, it was along toward evening and starting to snow. The stump rancher dumped our stuff and high-tailed it for home. We found a little hollow where we were sheltered from the wind and set up our tent facing a big rock. We built a fire against that rock and it reflected the heat so our tent was warm as a steam heated house. There were a lot of old windfalls, so there was plenty of firewood handy. When we got set up, it was snowing pretty good, so we piled a lot of brush around and over the tent, so that if there was a heavy snow, it wouldn't smash the tent. It snowed all night and all day for three or four more days. When it stopped, there was more than three feet of light, powdery snow on the ground. Then it turned cold. We had no thermometer to tell the temperature, but it was damn cold.

"There we were, snug as a couple of coons in a hollow tree. We were under three feet of snow, had plenty of soogans, plenty of firewood, plenty of grub. We even had fresh meat, for a nice spike (young deer), got curious the first morning and came too close. I shot him while sitting in the door of the tent, with no shoes on, toasting my feet by the fire. Later we got another without leaving the tent. Only one thing went wrong. One morning we forgot to warm the ax before starting to chop wood, and about half the blade snapped off at the first lick. It was a slow job chopping wood with only half an ax, but we had lots of time. Most of the time we just sat in the tent with the flaps turned back, a nice fire next to that rock out front, and played pinochle—life of Riley! I won $3,000,440 off of Don!

"After we got out, we heard that the company that sent us into the woods got worried because they hadn't heard from us. A clerk contacted old Sam and asked if he thought a searching party should be sent to look for us.

'Hell, no,' Sam said.

"But they have been gone more than two weeks and we haven't heard a word. They may be lost.'

'Them boys ain't lost. My boys don't get lost. They may be snowed in, or one of 'em may have broke a leg, or the crick might be up or sumpin, but they ain't lost. Just don't get your tail in a knot. They will come out when they get good and ready.'

"After about three weeks, we came to the conclusion we couldn't do any cruising till spring. It stayed cold; the only change in the snow was that there was more of it every few days. Anyhow, we were getting sort of tired of each other. Besides, Don said he wouldn't pay me that $3,000,440 because it was my deck and pretty badly worn, so he figured I could tell the cards by the worn spots, and that was cheating. Between you and me, he was right, though I wouldn't admit it. Of course, I never intended to press him to pay me—his wages were $90 a month and expenses—but it did feel good to have that much money coming.

"So we slung all the supplies we had left and our camping equipment in a tree to keep the varmints from getting 'em, and struck out. We got started about midnight. There was a full moon and a clear sky, and snow everywhere, so it was light as day. That was the longest 14 miles I ever walked. The snow was belly button deep, and so light and powdery, you went clear to the ground with every step. It was so powdery, I don' think we could have used snowshoes if we had had 'em, and we didn't. Never thought of needing snowshoes in that country so early in the fall.

"First I'd break trail for a few hundred feet, then Don would take a turn. If it hadn't been all down hill, I don't think we'd have made it. It was some better when we got to that old skid road, but not much. We got to that stump rancher's shack in the evening, just as it was getting dark. We had been 18 hours making that 14 miles, and if ever there were two tuckered boys, we were it. We had carried no lunch, for we had figured to walk out in five or six hours. The rancher's woman fixed us some grub, sowbelly and beans. I still remember how good it tasted. The rancher let us sleep on the floor in front of his fireplace and next morning hitched a team to a sled and took us toward Toledo. When we got down to the edge of the snow, we got another farmer with a wagon to take us to Toledo, where we caught the train.

"When we got to Portland, we went out to Sam's house. His wife let us in, and as soon as Sam saw who it was, he went to the phone, called the clerk who had been anxious about us and said, 'You remember that cruisin' team you thought was lost? Well, they're here. I told you my boys never get lost.' "

Louie also told me that once when he was carrying compass for Sam, they cruised a tract that was badly overripe. There were a lot of dead trees, and a lot of down timber. As they passed a big down tree, Sam saw something move in a little hollow under the log. He went over to investigate and found a little fawn, not yet a day old. He was looking at it, bending over a bit to see better, for it was sort of dark under the log, when Louie yelled to look out. He looked around and there was a big doe, evidently the mother of the fawn, reared up on her hind legs, and thrashing with her front feet, almost on him. Those flailing, sharp-edged hoofs could cut like knives. That big a deer could kill a man if she got him down. Evidently she thought he was a hazard to her baby, and she was not going to stand for it. He jumped sideways, just before those hoofs came down where he had been. Then the two of them left that vicinity as fast as they could pick 'em up and set 'em down. Sam would not go back to pick up the cruise where they had left off—he just entered in the cruise sheet what he thought should be there.

"That was the only time I ever saw Sam fake a cruise sheet," said Louie.

Schools and Kids

I served on the local school board for several years. One year we hired a new principal for our little high school. I had not known this fellow, but I had known his family, who lived in a neighboring community, and they were fine people. Ed seemed like a fine young man, just out of college, with a good presence, good scholastic record, an athlete, extra credits in school administration, a faithful attendant at church, etc. After school had been going on for a few weeks, we began to hear rumors of trouble. There were rather alarming incidents of misbehavior and impudence. The center of this trouble seemed to be Joe Smith, a smartalecky young punk, a sophomore, somewhat of a leader among the kids, who was always pulling sneaky things and egging on the other kids to violate school rules. Finally it was so open and blatant we had to take action. We decided Ed was woefully lax in discipline, but we didn't want to fire him outright during his first year as principal, as this might wreck his career as an educator. We decided to hire another teacher as an assistant to the principal.

A few miles away was a woman whom the members of the board knew. She had taught several years and we knew she got on well with the kids. This year she was not teaching because she had been needed at home. She was a big husky gal, who could do a man's work on the farm, but the family emergency had passed so she could take our job. She had not been on the job very long when Joe pulled one of his tricks and she caught him at it. Without saying a word, she picked him up, turned him upside down across her knees, and spanked him good, in front of the whole class. That solved the discipline problem. The kids had found out who was boss!

We were rather afraid the incident might cause trouble, for Joe's father was not an easy fellow to get along with. I met him a few days later and brought up the subject to get an idea of his reaction. His face lighted up with a big grin and he said, "That spanking was the best thing that ever happened to Joe. I've walloped him a lot of times for his impudence, but it never did any good. Now he doesn't even sass his mother. That teacher is O.K."

Several weeks later, when it was evident that we had no more discipline problem, I asked this teacher what was wrong with the principal that he couldn't handle the situation. She said, "The only thing wrong is that Ed was always a good little boy and just doesn't know how to deal with naughty little boys, but he is smart, he will learn; he will make a good school man."

She was right—he did!

A teacher in one of the small one-room country schools, a woman probably 40 years old, and not exactly beautiful, was working, grading papers after school one day, when she heard a commotion in the school yard. Two little boys were fighting. She got them by the collars, marched them inside, and gave them a lecture on keeping the peace. At the end she said, "Now kiss and make up."

They looked at her in astonishment, then at one another and one blurted out, "I wouldn't kiss him for anything."

"Me neither."

"You kiss one another, or you will both kiss me," said the teacher.

The kids looked at her, then at one another, then back at the teacher. It was evident she meant exactly what she had said. They slowly sidled together, then their heads turned and there was a token smack. As they left the room she overheard one boy say to the other, "I'd rather kiss you a hundred times."

Garibaldi took a lot of pride in its school, as I think was true of most small communities. School affairs were well supported, especially basketball. For a few years, we had a husky bunch of boys who played well, and we got into the championship playoff a few times, frequently with Cloverdale, in the south part of the county. They usually licked us.

In one game, with the score one point in our favor, and with only part of a minute to play, one of the guards fouled out. There was only one chap available to send in. Bob Simpson had played forward, and was a good one. He had been kept out of this game because of an injury. Since he was the only one of the squad left, the coach had to send him in, told him to stall for the few seconds left, just hang on to that one point lead. As soon as play was resumed, Bob got his hands on the ball, wheeled and pitched it through the nearby basket, and didn't realize until after the whistle blew that he had scored the winning point for the opponents.

Another year, the score was tied at the end of the second overtime period. In the third overtime period, Cloverdale made one point. Their team was heavier, our boys were tiring, and it was evident they were being outplayed. Most of this overtime period was played under the Cloverdale basket, but they were not able to get a field goal. In the final seconds, one of our guards, Roy Albers, got his hands on the ball. To get it out of danger for the remainder of the period, while facing the opponent's goal, with both hands he threw the ball over his head toward the other end of the court. The gun went off while the ball was in the air, but it went through our basket without touching the hoop. We came as near a riot that night as I have ever seen.

The funniest thing I ever saw depended so much on knowing the characters involved that I hesitate to make any attempt to describe it. However, it is worth a try.

Three people were involved. One was the clerk of the school district. Johnny was a Norwegian, a naturalized citizen. He had a good education and had worked several years in a Norwegian bank. He was an excellent clerk; observed all the details of the job and kept accurate records, though he had considerable difficulty with the English language.

The other two were elderly citizens, each having a little business in the community. Both were quite vocal and very penurious. They objected to everything that involved spending money. Especially did they object to spending money for schools. They considered themselves successful business men, and since all the schooling they had had was three or four years in the three R's, that should be enough for anybody. Neither had any small children or grandchildren, and they objected to paying taxes to educate other people's children. In short, they objected to taxes, period. Whenever there was any chance of the subject coming up, they were present, with long and loud speeches against it.

We needed a new high school building. The town was growing, so was the school population. The only way we could see to finance it was a bond issue, so the board voted to submit the proposal to the voters. It was up to the school clerk to write a caption for the ballot, describing the purpose of the proposed issue. As chairman of the school board that year, I was supposed to approve it before it was sent to the printer. Somehow that last little bit of red tape was missed, and when the ballot was printed, it was a very involved statement, actually stating what we wanted, but on casual reading, forbidding the board to issue bonds. It was just another illustration of Johnny's mixture of Norwegian syntax into the English language. Nobody on the board realized the ambiguity of the wording. Maybe we did not see a ballot before election day.

Much to our surprise, the two tax objectors worked for the bill. They went up and down the main street, buttonholing voters, escorting them to the polling place, talking to anyone that would listen about what a good thing it would be to vote 'yes' in that election. When the time came to count the ballots, as chairman of the board, I was present. So were the two old fellows. When the tally was announced, that the proposal had carried by a small margin, they both cheered and congratulated one another that the voters had had the good sense to clip the wings of the tax spenders. They felt good, and one asked the other to go to his home to have a little drink to celebrate. They had worked hard for that victory, but it was time well spent. Now was the time to celebrate. Until then, it hadn't dawned on us that the two old fellows had misread the ballot caption. We had wondered why the change of attitude, but were naive enough to think they had had a change of heart.

When it was explained to them just what the caption meant, that the bonds were authorized, but the margin by which they were authorized was very small, less than the number of votes they had brought in, that they had actually swung the election against themselves, they were too mad to swear. They just walked up and down the hall, each one talking to himself.

At last one of them stopped in front of Johnny, spit on the floor by his feet and said, "Damned Square-Head," and stomped out. We watched them walk up the street about a hundred feet apart. They were too mad even to talk to one another. I doubt if they had that drink.

You can't always tell by the school record how a kid will turn out. One boy in the county was thought to be stupid. He never had his lesson; he didn't seem to care whether he ever got it or not. He paid no attention to what the teacher said or to anything else in the schoolroom, except the clock. Finally he dropped out.

He got a job in an electrical shop and just seemed to come alive. He appeared to know instinctively how to put things together so they would work, or how to fix things that didn't work. He went on to a bigger shop and soon he was in charge. He invented a series of gadgets that seemed to be needed, and now, I am told, he is a millionaire and has a responsible job in a big organization.

You can't always, sometimes, tell.

The Boy Scout movement came to Tillamook early, and there were several strong troops in the county, which was designated a district. Three or four times a year there would be a district Court of Honor, where aspirants for merit badges would demonstrate their knowledge of the subject under consideration. One such meeting was held in the Garibaldi schoolhouse. There were several candidates for the Life Saving badge, and I was appointed examiner. We went over the theory part of the examination, then I had each candidate choose a boy for a patient, and strung them around the room to show their stuff, each boy doing one thing.

Bob Berkey, a sharp lad from Tillamook, was demonstrating artificial respiration, and doing it well. After watching him awhile and asking why he was doing what he did, I asked, "Bob, how long would you keep this up?"

"Until I am sure he is dead," was the immediate reply.

The Cub program came later. I dropped in on one pack meeting where the leader was a man recently discharged from the Navy. He was a cocky young fellow. When I got there, the boys were gathered around him, asking questions about Navy life. I stood quietly in the doorway, unnoticed. He was obviously pleased to have them so interested in what he was telling them. I knew nothing of Navy life, but from a little experience in the Army, I thought he might be stretching things a bit. One lad asked, "In the Navy, can they make you do things you don't want to do?"

"Naw, they can't make you do nothin' if you don't want to do it," was the reply. Then he caught my eye, probably saw a bit of questioning there, and added hastily, "But they sure can make you wish you had done it."

I guess that is the essence of discipline.

A child's viewpoint is hard to get at times, and sometimes when we get it, it is a shock. Thomas Hood said:

I remember, I remember;
The fir trees dark and high;
I used to think their slender tops
Were close against the sky;
It was a childish ignorance,
But now 'tis little joy
To know I'm farther off from heaven
Than when I was a boy.

Sally was maybe four or five years old, daughter of a close neighbor and friend. It was a common joke to refer to her awkwardness. She would trip over her own feet, and was always running into things and hurting herself. She was a dreamer type of child, who lived in a world of her own, and we thought she just didn't pay attention to what she was doing or where she was going.

A long time after I should have had the idea, it dawned on me that maybe this child could not see as well as she should. So, her parents took her to Portland for examination and a visual deficiency was discovered. Glasses were prescribed and ordered. For some reason, when they were finished, the optician sent them to me. That evening I delivered them and put them on the child. When I was ready to go, the family walked out on the porch with me. It was a beautiful summer evening; the stars as brilliant as you see them only when you are far away from city lights and smoke. When she came out, Sally let out a squeal of excitement.

"Oh! Oh! Oh! Oh! look at all the holes in heaven!"

Then it hit me like a club; that child had never seen the stars before. And what else had she missed? How many bumps and falls had she taken just because my thinking machine needed a grease job!

The confidence of a child is sometimes humbling. A widow with two small boys, maybe five and six, lived next to the office. They were well behaved little fellows and I would sometimes take one or both when making a trip. One day one of them was riding with me when I had a small accident. I was driving too fast for conditions, or not paying enough attention to what I was doing, or something. Anyhow, the car left the road and mashed down some brush. I was able to get it back on the road without stopping, and with no damage beyond scratched paint. A little further off the road would have meant a bad wreck. The boy was too little to appreciate the danger. He thought it was a real adventure and was thrilled. I was ashamed of my carelessness, especially when he was in the car. The mother heard us drive up and came out to call the lad in to lunch. He couldn't wait to tell her about the incident. When he had finished she said, "Weren't you scared?" He looked at her in wide-eyed astonishment and said, "Why no. Doctor was driving."

Kids are kids, anywhere. It was a smart guy that invented them. You never know what they will say or do, but usually it makes sense if you will just try to see things as they do.

One spring day, on the last day of school, a cute little girl, about eight years old, showed up at the office. She had been out of school for a few days because of some minor illness and she wanted a permit to return. If she left the office at once and walked steadily, she could not possibly get to the schoolhouse more than fifteen minutes before dismissal time, so I could not see any use bothering to fill out a permit, and told her so.

"But I gotta have it. Teacher won't let me in without it and I wanna get back in time for vacation."

Jimmy Smith was a high strung kid, of more than average intelligence, about eleven years old. He had a severe illness, in the course of which he became quite depressed and severely dehydrated. I was not sure why this happened, but thought it largely due to an emotional rejection of fluids. He would heave every time he took a drink. While talking with him one day, I discovered he was very fond of soda-pop, which his mother allowed him very sparingly because she thought it was not good for him. We had quite a lengthy discussion of the various flavors. After talking about each, he would dismiss it with, "But Mom won't let me have that kind; says it is not good for me."

Then I said, "Jimmy, if you had all the soda-pop you wanted, how much do you think you could drink in a day?"

The prospect was dazzling. All he could drink! His eyes shone as he said, "I could drink a whole case" (24 bottles).

I argued the point with him, but the more I argued, the more positive he got.

At last he said, "I'll bet you a nickle I can drink a whole case in a day."

I still argued, but let him talk me into a bet that he could do it in two days. He thought that was easy money, and was jubilant. He showed more interest than he had for a long time. I told him I would tell his dad to go to the store and get it.

I still had a selling job to do on the mother. When I got her where Jimmy could not hear, I gave her a long song and dance about the boy's acidosis and dehydration and how he needed fluids, and how the alkali in the carbonization of the soda-pop might be just what he needed, etc., etc. Finally she gave grudging assent. She had always been told that soda-pop was bad for growing children, but conceded it might be used as medicine.

Two days later, when Jimmy saw me walk in, he held out his hand with a sticky nickle. Tears came into his eyes, and he said, "I just couldn't do it."

Inspection of the soda-pop case revealed one full bottle and about an ounce in another. If you don't think that was a lot of soda-pop for one small boy, try it. Whether the soda-pop did it or not, Jimmy started to get better; his mother was convinced, and he stayed on a heavy soda-pop diet for some time.

Ever experienced motion sickness? I have been seasick, plane sick, car sick; I can even get sick on a roller coaster, Ferris wheel, or a swing, much to the disgust of my kids.

One summer, we had some small nieces and nephews visiting at our home. The captain of a tugboat invited me to bring the whole bunch for a ride with him. When we got out in the ocean where we were to meet a ship to be brought over the bar, it was not there. Somebody goofed in making the arrangements, and it did not show up for an hour. The captain ordered the anchor dropped and the engines stopped. I had been a bit uneasy crossing the bar, but when the boat started the combined roll and pitch of riding at anchor, I became very uncomfortable. It took some effort, but I was holding my breakfast.

In the meantime, the kids were having a ball. They didn't know they were supposed to get seasick. The crew had little to do while the boat was anchored, so entertained the kids. Among other things, the cook took them to the galley and fed them. When they came up on deck, there was jam on their faces and their hands were full of cookies and fruit. One of the boys came running up to me, held up the fruit he was eating and said, "Unca Doc, want

The author and family. One boy missing; also dog.

part of my banana?"

That did it. I barely made it to the rail!

During the hard times after the market crash of 1929, a lot of people had a really tough time, and we had our share of them. One group of six or eight families got together and tried to make a living cutting firewood. They built shacks of hand split shakes and bark, in the edge of a pretty scrubby patch of timber. They had no money for machinery, so they were doing everything by just plain brute strength. One day when I drove past the place, my boy, then eight or nine years old, was riding with me. It was a miserable day, dark, cold, raining hard, windy, and had been for several days. These people, both men and women, were out in that rain, without proper clothing, soaked to the skin, in mud halfway to their knees, trying to man-handle poor grade logs into something salable. It was a most depressing sight, and one I have never forgotten. Just to relieve the impact of such a sight, I said the first thing that came to mind. "Son, if worse came to worst, do you think you and I could take an ax and saw and make a living for the family?"

There was no immediate reply, and I thought he considered the question too silly to answer, for I knew good and well we couldn't do it and thought he knew it too. A couple of miles down the road he said, slowly, "I don't know. I guess we could do it if we had to, but it sure would be tough." Another pause, then, "But I betcha we'd find a better way to do it."

That struck me as a typical American attitude. There is a better way to do it, no matter what the problem, and we are going to find it.

Kids have an innate sense of fairness that is sometimes surprising. For a time we had a teacher that was very unpopular with the kids, I do not remember why. They always referred to her as "The Old Soupbone," which name might give some clue to her status with them.

Late one evening, long after time for school to be out, I was passing the schoolhouse, and saw a boy whom I knew coming out. I stopped and spoke to him. "Pretty late to be getting out of school, isn't it?"

"I had to stay in."

"How come?"

"Oh, a bunch of us were picking on the Old Soupbone and I got caught."

And he went whistling off down the street, with not the least bit of anger or resentment in his mind. He got caught and deserved the punishment. Punishment for being so clumsy as to get caught, I mean. What he got caught at was perfectly legitimate.

Churches, Events and Such

There were two small churches in Garibaldi. As was true of most churches in small towns, both of these churches had a hard time. Membership was small, money was scarce, everything had to be built from the ground up. As in all churches, those who were loyal to it and gave it their time and work, as well as their money, developed a respect for and appreciation of one another that would be unusual in any other sort of organization. Class lines were erased; everyone met on the same level; all were working for the same end. We made friendships in our little church that have persisted over the years. Some were people we would have known only casually, or professionally, but never personally.

One of the impressive things about these small churches was the ministers. They were not brilliant men; they were not outstanding speakers, or deep thinkers, but they were earnest and sincere, and really wanted to be of help. They worked for salaries that were woefully small. Most of them I do not remember by name. One I do remember was John Seethoff, a German, who taught me a lesson in tolerance I have never forgotten.

He was a good man and a fair preacher, but made some unusual uses of the English language. One time there arose some sort of difference of opinion in the church. I don't remember what it was about, but feelings became rather heated. Finally, John announced that he would preach on that subject the next Sunday. There was a full house. He took the entire time explaining the arguments of the group opposing him, and ended up with, "And they may be right. I am not real sure."

For some reason, nobody seemed to care about the argument any more.

In a neighboring community, a small church was trying to raise money to pay for a building, and were having a hard time of it. In fact, it looked like they had bitten off more than they could chew, and would lose what they had put into it. Their plight came to the attention of a professional gambler, who promised to help. In a few days it was pay day for the mill and camps. He started a poker game in the rear of a tavern, carried it straight through for 48 hours, and turned all his winnings over to the little church, enough to clear their building. I have wondered if they deleted gambling from their list of sins.

One of the Tillamook churches had a piano to accompany the singing. It was located on the main floor, just in front of and to one side of the platform. The pianist was facing the choir leader, with her back to the audience. One Sunday, while the choir was singing an anthem, the pianist, and those in a few front pews, could see a mouse come out from under the pulpit and play around on the floor, not far from the pianist's feet. It was a good thing the pianist was familiar with that anthem; she spent more time watching the mouse than she did watching her music. The mouse was having a good time scampering here and there, investigating all sorts of little things that seem to interest a mouse. Sometimes it was only a few inches from the pianist's feet; then it would run back. This went on for several minutes. Then came a time of relative quiet; the choir sang in a low voice, and the piano tones were muted. This ended with a sudden volume of sound. The choir came out at the top of their voices; the pianist stepped on the loud pedal and banged the keys with both hands. The sudden noise alarmed the little beastie and he scurried out of sight.

Later the pianist said: "Sure I was scared—I'm afraid of mice. I picked out a certain crack in the floor and decided if that animal crossed that crack, I was leaving, anthem or no anthem. He came up to it four times, and once poked his head over. I was just ready to jump, but each time he went back. I don't remember playing any of the anthem, but the leader said I never missed a note. Guess that is the reward for all the time I put in practicing."

There was a little Sunday School in a one room schoolhouse in the south end of the county in the early days. They had an old fashioned foot-pump organ and a very faithful organist. Her only shortcoming was that she knew only one song and couldn't read music. The superintendent said, "She played that song real good and the children sang it real good, too. They all knew all the words to all the verses. But it was kinda discouragin' having to sing the same song over and over, maybe four or five times each Sunday, and the same thing every Sunday."

Courts

In a rural community there is always interest in court procedures. A few cases will be long remembered.

One concerned a prominent citizen, getting along in years. His wife was dead and he had a housekeeper who lived in the house. Due to the infirmities of advancing age, he had to spend considerable time in a wheelchair. So that he might get in and out more easily, he had a ramp built to his front porch. When it was completed, he and the housekeeper had an argument over who would try it first. The housekeeper won the argument. She got in the wheelchair, started down the ramp, ran off the side, tipped over and broke her arm.

She wanted compensation for the injury, claiming it was incurred in the line of duty. He refused; claiming riding around in his wheelchair was not what she was being paid to do; she was just horsing around. She sued him and the case came to trial. I do not remember for sure how it came out, but I think she collected.

A fellow bought a pair of shoes from a local merchant who had the reputation of being a bit shifty. In three days the shoes came apart; the buyer took them back and demanded a refund. The merchant refused, claiming he had made no guarantee of the shoes, therefore was not liable for their performance. The buyer was not about to lose the price of a pair of shoes. He was so persistent and raised such a disturbance the merchant phoned for the marshal and lodged a complaint of drunk and disorderly. When the case came up for trial, the purchaser introduced the shoes as evidence and established by several witnesses that he was a teetotaler, also that he had driven a school bus for several years, which was strictly a job for a non-drinking man. Further, he proved that he had for many years been an active member and officer in a religious group that strictly forbids the use of tobacco or alcohol. Of course he was acquitted.

Then he filed a suit against the merchant for defamation of character and collected $1,400. There wasn't much profit for the merchant in that pair of shoes!

An ambitious citizen thought Tillamook needed a wine shop so started one. He was probably his own best customer. In fact, he was about his only customer. Tillamook people did not seem to go for wine. Those who wanted liquor seemed to want something more potent. At last conditions got so bad that there was a fire in the store, but the firemen arrived promptly and put it out.

The owner filed a claim for insurance, listing his inventory. When it came to trial it was proved that most of the bottles that were not broken by the fire were only half or a third full. His claim was not allowed in full, but they were not able to prove arson.

A well known alcoholic was arrested for driving while drunk. He pleaded guilty. The judge announced that since he had pleaded guilty and eliminated the necessity for a trial, he would not be fined, only required to surrender his driver's license. He reached in his pocket and handed the judge his liquor license. The judge was not satisfied with that, but investigation showed that his driver's license had expired two years before and he had not renewed it. He claimed, and was supported by witnesses, that twice during that time, when caught in routine road blocks, he had showed that liquor license to the police officer, who had glanced at it and waved him on. Therefore, he claimed, that was his driver's license because he had been authorized by police officers to proceed when he showed it. He had offered it when the judge ordered him to do so, and so had fulfilled the penalty imposed on him by the court. The court refused to accept that and imposed a fine. The case was appealed to a higher court and the lower court was overruled.

County Fair

One of the main attractions of the Tillamook County Fair was the pig and Ford race. This race was run with Model T Fords, stripped down to the springs, and with a flat platform bed. No gadgets were allowed, no seats or starters. The pigs were vigorous shoats, weighing about 30 pounds, three pigs for each contestant. They were in a pen alongside the race track. The cars were lined up at the starting line with the motors stopped. At the starter's signal, each driver ran to the pen, caught a pig with his bare hands, carried it to the car, held it with one hand while cranking the Ford, then drove around the track, driving with one hand and holding the pig with the other, stopped the motor and deposited the pig in a second pen. Then he ran back to the first pen and repeated the performance, then did it again. The first one to make three rounds of the track, each time with a different pig, then to get his car to the finish line, was the winner. If a pig got away, it was necessary to stop the car and run it down. Holding a squealing, kicking porker with one hand while cranking and driving a jalopy with the other is no easy stunt. Many were the mishaps and many were the laughs.

The race is still being put on and each fellow owning an old Model T takes better care of it than he does of his new car. There are not many of those Model T's around, and most of those are owned by museum or antique car buffs, but there are at least six of them in Tillamook County that are still working cars. They are not only used at the Tillamook fairs, but are in considerable demand at other fairs.

Other features of the Fair were horse pulling contests and logger sports. In the pulling contests, which are put on in many fairs, the teams were classed by weight. Each team in turn was hitched to a sled, weighted to a certain limit, depending on the class. The team was then given a chance to pull, and if they could move the sled a stated distance, the load was increased until they could not move it. The varying techniques of the drivers and the responses of the teams made an interesting spectacle.

In the logger sports, such things as chopping, sawing, tree topping, things the loggers did in their daily work, were done against the stop watch. Each camp would enter a team, and the results were scored something like a track meet. Competition was keen and the rooting sections were as partisan and as noisy as those at intercollegiate competition.

Of course there were exhibits of livestock, cooking, handiwork, vegetables, flowers, etc. As you would expect in a dairy county, the exhibits of dairy cattle were extensive and good. Some claimed the Tillamook County Fair surpassed the State Fair in this department.

The first Fair was held in 1925, and the 40th in 1966. There was a gap during the war.

Clubs and Miscellaneous

For many years, Tillamook had two service clubs, Lions and Kiwanis. There was much rivalry, each trying to outdo the other. Sometime in the twenties, the Lions held their state convention in

"First Day of School" cast, c. 1924, play put on for benefit of Garibaldi Methodist Church. Memorable to author, front row (right end) because another player's wild swing with dinner pail broke and bloodied his nose, which audience accepted as part of play.

Tillamook. Advance registration was about 200. The opening session was to be a banquet in the evening, at which Gov. I. L. Patterson was the principal speaker. Elaborate preparations for this banquet were made. The largest hall in town was rented; music, novelty numbers, decorations, etc., were prepared. The women of one of the churches were engaged to prepare and serve a sumptuous meal. They were guaranteed a certain number of plates. There had been no specific reservations or ticket sale for the banquet, the assumption being that most of the delegates registering for the convention and their wives would attend. That had been true in previous years when the conventions had been out in the Valley, within easy driving distance of most of the clubs. What was different this time, was the long drive across the mountains. Most of the delegates found it inconvenient to arrive in time for the banquet. An hour before banquet time, only a handful of delegates were on hand. It looked as if Governor Patterson would speak to the local Lions, their wives and a large assemblage of empty chairs. And all that good food had to be paid for!

At the last minute, a general invitation and call for help was issued to the Kiwanis Club and American Legion. They responded nobly. I do not know whether the prevailing motive was to keep

Tillamook from the disgrace of ignoring Oregon's governor, or the appeal of a free meal, but they were there in force. The tables were filled, the food was eaten, the governor and the rest of the program was listened to and applauded, and everybody had a good time. I think members of the Lions' Club enjoyed it less than the others, for they knew that their club was stuck for that feed. They had guaranteed the bill, and the club didn't have that much money in the treasury.

Next morning, the officers signed a note at the bank and got the money to pay the bills. Then the District Governor of the Lions told the story at the session next day; told the delegates they were responsible because they had not shown up, and asked them to pay for the banquet, even if they had not eaten it. Most of them did, enough that the note at the bank was cleaned up. But the Kiwanians remembered for years that they got a free meal from the Lions.

Before there were bridges over all the streams too deep to ford, there were ferries. It was a part time job for a nearby settler and brought in a little cash money. On the Nehalem, Otto Effenberger, then in his teens, operated such a ferry near their home. The fare was five cents. Anybody wanting to cross from their side would come to their house. On the opposite side, he had hung an old wagon tire, and anyone wanting to cross would bang on that to attract attention. One day he was doing something he could not very well leave, when he heard someone banging on that tire and called to his brother, "Hey, Oscar, there is a nickel on the other side of the river. Do you want to go get it?"

It happened that the prospective customer was a very pompous, self important man. He heard the call and he was terribly insulted to be referred to as a nickel.

When Mr. Effenberger, Otto's father, built his first cabin near the Nehalem River, he made a doorstep by dressing off one side of a large boulder. He left coarse tool marks on this rock so it would not be slippery when wet. When the cabin was finally torn down, the large stone was in the way, so it was rolled back to the river bank, where it laid for many years, gradually becoming almost buried in silt.

One day an amateur archeologist came by, fishing. He saw the old tool marks on this stone, thought they were pre-Indian hieroglyphics, and became much excited over his find. He expended a great deal of time and labor getting the boulder up to the road and loaded on a truck. Nobody ever heard whether he succeeded in deciphering those tool marks!

A farmer named Moore was a cantankerous chap; had trouble with nearly everybody. He had land on both sides of the river, near the bay, where the stream was somewhat affected by the tide. There was a gravel bar where a man on horseback could ford the stream. When he wanted to work across the river, he would jump on a horse and ride across at this place.

One day he fell off his mount and was drowned. Nobody saw what happened, but probably the horse stepped on a loose stone and fell. It happened just as an unusually high tide was coming in. The body was found some distance upstream. His teenage son was helping pull the body from the water when he observed, "Dad is still as contrary as ever. He floated upstream."

Ezra Thomas was a real neurasthenic. He had every disease he read about in the patent medicine almanacs, and never a day passed that he didn't take several doses of some kind of medicine. He and his boys were cutting hay at some distance from his home, across a sizable creek. One day they had a wagon loaded with hay and he insisted on taking it home. The boys objected; said the creek was too high to ford, but he climbed on the load and headed toward the creek. Sure enough, when he got in midstream, the water piled up above the wagon and tipped it over. The team scrambled out, but Ezra couldn't swim. He managed to get hold of an overhanging limb and hung on until the boys could get to him and pull him out, but he was in the cold water long enough to get thoroughly chilled.

They took him to a nearby farm house, where they wrapped him in blankets and built up a good fire in the fireplace. He was still shivering, and had not gotten over his scare. "I need some medicine, need some medicine. Mrs. Tate, haven't you got some medicine I can take?"

The Tates were not medicine addicts, but she looked around and found a bottle of Watkins' liniment. She poured out a generous slug, filled the dipper with water, and he downed it.

It did him a lot of good!

Isolation and Transportation

The railroad was built into Tillamook County about ten years before my time, and was the principal route into and out of the county until the hard surfaced roads were built. Boat service was unable to meet the railroad competition and had given up the struggle. There was a passenger train daily, each way, and while it was only 99 miles from Garibaldi to Portland, it was an all-day trip, and a very interesting one, too. It passed through much virgin timber, over some very rough country, and was always subject to delays from weather, slides, fallen trees, etc. There were 13 tunnels, the longest one 1,437 feet. There were 35 bridges over 100 feet long. One bridge was 167 feet above a canyon, the highest bridge ever built with single span piling. This bridge was burned in the big fire of 1933, and piling long enough to rebuild it was no longer available, so the new bridge was built in a more conventional manner.

A trip over this road was as much a social event as a journey. Except on summer weekends, the cars were not crowded; people usually knew one another, or soon became acquainted. The train stopped at every village and many of the settlers' houses, to drop off mail and supplies. The people all turned out to see the train go by, and came into the cars to see if any friends were aboard, if the stop was long enough. If there was a slide or tree across the track, most of the men would get out to help clear if off. Everybody carried a lunch, and when noon came it was like a picnic. Nobody was in a hurry, not even the train crew. The schedule was pretty generous, allowing for delays, and if they didn't make the schedule nobody cared very much.

The railroad was more than a way in and out of the county. Many railroad employees considered themselves citizens of the community, were active in local affairs and were always available to help when needed. When the only north-south road was impassable, as was frequently the case during winter storms, I made many a trip on a handcar, one of those "come-to-me, go-from-me"

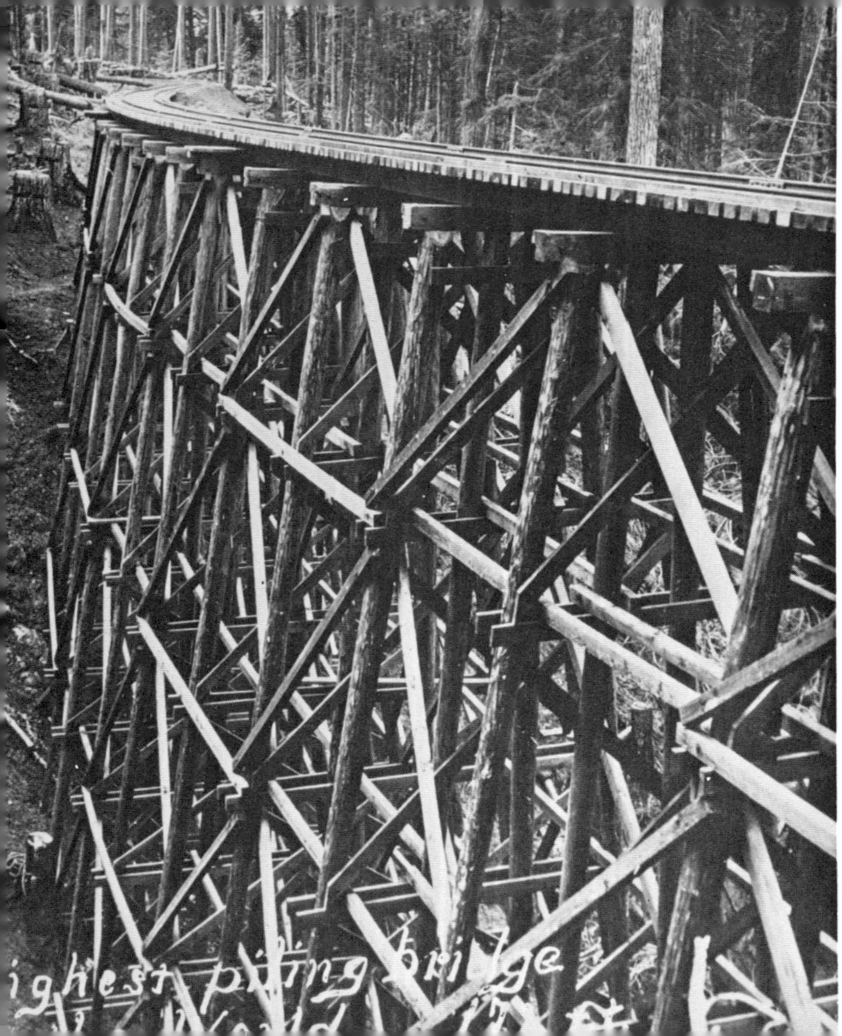
ighest piling bridge

rigs, propelled by a couple of husky section hands, who would accept no pay for their extra work.

When the automobile roads were good enough to drive out in a few hours, railroad passenger service was no longer needed and was discontinued. Now long freight trains are the only traffic on the railroad. I guess that's progress, but sometimes I wonder which way? It is probably more profitable for Southern Pacific without the passenger traffic.

The railroad came to the coast area down the south fork of the Nehalem River, then along Nehalem Bay to the ocean, then south along the coast to Tillamook. That was a natural water grade. The north-south highway parallels the railroad from Tillamook to the mouth of the Nehalem River. For many years there was a gap of about three-fourths of a mile between Nedonna and Brighton, where there was some rough ground, and the county had never gone to the expense of making a road. To go around by the inland road was about 15 miles farther. Most people left their cars at the end of the road and walked the railroad across the gap, hoping to pick up a ride to their destination on the other side.

There was a constant temptation to run an automobile on the railroad right-of-way. Running on the cross ties was pretty rough going, but not much worse than most of the road. There was only one passenger train each way, but a freight might come along any time. The road was so crooked you couldn't see from one end of the stretch to the other, and the woods were so heavy you frequently couldn't hear a train a half-mile away unless the engineer happened to blow the whistle to scare a deer or bear off the track. Also, it was strictly forbidden to drive on the right-of-way, so it was a real sporting challenge, and one that was frequently accepted.

One night a bunch of young fellows at Brighton decided to go to a dance at Bar View. It was too late to go the long way around. At that time of day, the chance of picking up a ride on the other side of the gap was very poor, so they decided to take a chance and drive on the railroad right-of-way. When they were about midway of the gap, they heard a train whistle, apparently too close to allow them to reach the other end. They had to get off the track. That was no problem—five husky mill hands just picked up that Model T and set it alongside the rails. But in the darkness they

Left, horse and buggy thrills, Neahkahnie Mt., c. 1910 (OHS).
Right, highest single-span piling bridge, Pacific Ry. &
Navig. Co. (SP) line to Tillamook. (Courtesy G. B. Abdill)

The railroad along lower Tillamook Bay was mostly on piling or fill. Buildings along it were on piling, over swampy ground. Gradually, the area behind the railroad was filled. The two circa 1920 photos show Garibaldi and Barview depots. (Wesley Andrews Collection, OHS)

set it too far over and it rolled over the bank and down on the beach. Fortunately it was low tide, and a little thing like a ten-foot drop wouldn't hurt a Model T, but that car had to be out of there before the tide came in. After the train passed, a couple of the boys hiked back to the mill, got a strong block and tackle from the shop, took a tailhold on a tree on the opposite side of the right-of-way, a few good heaves and the car was back on the railroad track.

They were late for the dance, but not as late as they would have been if they had gone the long way around. Their good clothes were a bit mussed and spotted up, but that was a small matter—they had a good time at the dance. They returned home the way they went and didn't have to give way for a train.

Transportation in Tillamook County was a constant problem. We were trying to use automobiles on roads made for horses. Any major change in a community's customs is bound to produce interesting incidents. The change from horseflesh to gasoline was no exception. I presume the transition in our community was no different from that in others, but I only knew of ours.

The machines themselves were crude and undependable. Breakdowns were frequent, but relatively simple to fix. The usual mechanic was fairly well equipped if he had pliers and screwdriver, in two or three sizes, a crescent wrench, a file, a hacksaw and a hammer. Baling wire, cotter keys, bolts and nuts were the principal items in the parts cabinet. Adjustments were made by eye and ear.

Things we take for granted were extras or unheard of. Windshield and tops for roadsters or touring cars were extra. Horns were operated by a hand bulb. Windshield wipers were seldom seen, and were manually operated. You carried a stick to check the amount of gas in your tank. The first tops for cars had oilcloth side-curtains that were put on with turnbuttons. Closed cars, coupes and sedans, were slow to appear, and for a time their use was evidence of effeminacy—only sissies couldn't take the weather.

Tires were a constant source of trouble. They were the tube and casing type and their life expectancy was short. When the first guaranteed tires came on the market—Royals, I think—guaranteed for 10,000 miles, many of us refused to buy them. We thought it was a gyp; no tire could possibly go that far. Most cars carried two spares and a tire patching kit. You just expected to have a flat every few miles. With two spares you might get home, if you didn't go too far, but frequently you didn't. When you had a flat after you had used your spares, you just pulled off the road, took the rim off the wheel, and the tire off the rim, patched the hole, put the tube back in the tire, the tire on the rim, the rim on the wheel. Then you pumped up the tire with a hand pump. On a trip of ten miles, it was customary to see three or four cars undergoing roadside tire changes or repairs. You never got in your car before you walked around it to inspect the tires, kicking any that looked doubtful. I don't know that kicking a tire did any good, but you did it anyhow.

Because of tire trouble, rough roads, hard steering cars, mud-holes, etc., very few women drove, but there were a few hardy souls that braved the risks.

There were no such things as driver training or licensing. Anybody who had enough money to buy a car was supposed to have sense enough to drive it. When you bought a car, you got a few minutes instruction in what the various knobs, levers and pedals did. If he was big-hearted, the salesman would sit beside you for the first few miles. Even so, there were surprisingly few accidents, probably due to slow speeds. Forty miles an hour was a dizzy speed, attained only by the best cars on the best roads. The average was probably half that, or less.

One fellow told me: "One day I was standing in front of Lamar's saloon when Ed Hadley, the Dodge salesman, drove up. 'Come take a ride,' he said, and as I had about an hour to spend before milking time, I got in. We rode around maybe a couple of miles, ending up in front of Ed's garage. We sat in the car for a few minutes, talking of various things. After a while he said, 'What do you think of her?'

'Pretty nice,' I said.

He named the price.

'O.K. I'll take it.'

'It's all yours; pay me when you get around to it,' he said, then climbed out and went into the garage.

"I slid over into the driver's seat and tried to remember what I'd seen him do. The motor was still running, and I finally figured out how to get it in gear and headed for home. The last half mile of road was grown up so heavy with brush there was just room for a wagon to get through. That was pretty tight for a car. When I had gone a little way on this stretch of road, I came around a big stump, and there was my neighbor walking towards me in the middle of the road. He was so surprised he just stood still; anyway, the brush was so thick I don't think he could have gotten out of the way if he had wanted to. I couldn't think how to stop the dad-blamed machine; thought I was going to run over him, sure. Of course I wasn't going very fast—I found out later I had it in low gear—but it seemed to me I was just tearing along, and I guess it looked that way to him, too. Just as I got to him, he came to, jumped, and landed right on top of the hood, where he sat until I could figure out how to stop.

"That is the nearest I ever came to having an accident, and I have been driving a car ever since."

The Model T Fords had a pedal you pushed on to keep it in low gear, giving rise to the definition, "A Ford is a car you push up hill with your left foot." Several instances were known where a man who had driven horses all his life would plant his feet firmly, one on the low gear pedal, one on the gas, pull back on the steering wheel and yell, "Whoa" when he wanted to stop. With that combination, of course, the car didn't "Whoa," but went through any fence, rose bush, chicken house or shed that happened to be in the way. Stumps or trees were different; they stopped it. Probably the most frequent automobile injury then was a broken arm. There were no starters, and you had to adjust the spark just right before you began to turn the crank or the explosion would come too soon, the crank would be jerked out of your hand, spin around and strike the upper third of the forearm.

About the middle of the twenties, I think, a very small car was marketed in our area. I do not remember the name. One was bought by a fellow in Garibaldi, a tall, skinny guy named John Young. It was funny to see him folded up in that car. He didn't have it long before he wrecked it. The common explanation was: "John pushed on a pedal so hard his foot went through the floorboard and he tripped the car."

One day, while crossing the mountains on the way to Portland, my car quit. A passerby towed me to the next garage. The mechanic came out; I told him what happened. He opened the hood, hit the vacuum tank with his fist and said, "Now try it," and the

Single-lane driving adventure in Coast Range, late 1920s. (OHS)

motor took off like a sewing machine.

Then he explained to me that the cork float sometimes stuck to the wall of the vacuum tank. When I asked his charge he said, "Oh, about a nickle for hitting it and forty-five cents for knowing where to hit." That sounded reasonable so I paid him and started out. A few miles further on, the motor died again. This time I tried hitting the vacuum tank and got results. This happened every few miles all the way to Portland. When I started home, it was the same story. Finally I got tired of the game. I cut a husky switch, drove with the hood open (cars then had the hood hinged along the center, not at the base of the windshield as it is now, so that each side opened independently without materially obstructing the view), and every time the motor stopped I'd lean out the window and give that vacuum tank a vigorous whack with the switch. The mechanic at home said it was the only time he ever heard of using a whip to drive a car.

Troubles were so frequent everybody was willing to help everybody else. I always carried a piece of good stout rope, for there was no telling when I would find somebody stuck in a mudhole. The chances were better than fifty-fifty that that someone would be me. Only once did I ever know of anyone refusing to help, and never did I hear of anyone taking pay for helping.

A neighbor told me one day that he was out near Willamina, where he saw a fellow working on an old beat-up Ford alongside the road. The man had on old clothes and he was muddy and greasy to his ears. "I pulled over and asked if he needed any help. He said he had a broken axle, and would appreciate a ride into Willamina to get a new one. So I took him in to Willamina where he found a new axle and borrowed some tools; then I took him back to his car. I offered to help put in that new axle, but he refused, saying he could manage by himself.

"Three days later, I had to go to Otis. The road down that way was pretty rocky, and before I was anywhere close to Otis, I had two flat tires. That used up my spares, and I was hoping I could make it to the Otis garage. But no such luck. Pretty soon there was a bang, followed by a 'siz-z-z,' and I was bumping. I had used up all my patching material the day before, and had forgotten to get more before I started, so I was in a fix. As I was standing there looking at that flat, and trying to make up my mind to start walking, a nice shiny Buick drove up, stopped and the driver got out. He saw my problem and offered help. I asked for patching material, but he didn't have any either. He offered to take me to Otis. We took the two flat spares off the rack, went to the Otis garage and waited while the mechanic fixed my tires. My rescuer was a well dressed fellow; seemed to be pretty well fixed, at least he was driving an expensive car and he had all the extras on it that I had ever heard of. He looked sort of familiar but I couldn't place him. I kept looking at him as we were waiting, and at last I said, 'You remind me of somebody, but I can't remember who. Do I know you?'

'You ought to. I'm the fellow you hauled around Willamina last Tuesday!' "

I soon found out that most cars just wouldn't hang together under the punishment. After trying several makes, I found the old Hudson. Those cars could take it. They just didn't shake apart, and were heavy enough to get traction on steep slopes, along with power enough to climb hills or plow through deep mud. I wore out, or wrecked, seven of them before they went off the market. I could not understand why they were not more popular.

One other advantage of the old Hudson — it would wade through deep water, because the carburetor intake and battery were so high on the engine. If you disconnect the fan belt, so the fan does not splash water on the spark plugs, and drive slowly, so the water can flow smoothly around the car, any car will continue to function until you get water in the carburetor, or short out the electrical system. In the Hudsons, I have gone through water so deep I would be wet halfway to the knees as I sat with feet on the pedals. Of course you have no brakes when you get out of the water, but that is easily remedied — just drive a short distance with your foot pressing the brake pedal. Heat from friction will soon dry the brake lining.

The first four Hudsons were the coach models, two-seat, two-door jobs. I would have the back of the right front seat worked over so it would lie back, and I had a sort of bed to carry a patient on a stretcher. I had a folding stretcher that could be carried in the trunk. The only ambulance service we had was the Tillamook undertaker with his hearse, and sometimes it took him a long time to arrive; he might be out when we called. Besides, for some reason, some people just didn't like to ride in the hearse. So my makeshift ambulance, which was always with me, was used frequently.

Harry Mitchell was a good, honest, dependable Indian, a fisherman who worked steadily and made a good living. His wife, a white woman, had multiple hernias that defied all efforts to correct. She frequently got a strangulation that required general anaesthesia to relieve. She made so many trips in my ambulance that one day Harry propositioned me to buy the car; said he thought it would be cheaper to own it than to pay me to haul his wife around.

Once I went out on a lonely road to pick up a man found in the ditch. As far as I could see, he was just plain dead drunk. The only abnormal thing was that he was barefooted. We loaded him into my improvised ambulance and started toward his home. After a time, for no apparent reason, the motor stopped. I fooled around with the engine a bit, and it started. Soon we went through the whole procedure again. That happened three or four times before I discovered that my supposedly unconscious passenger had waked up and was turning the key with his toes. He thought it was funny, but I didn't and tied up his toes with adhesive tape.

One night, driving a lonely road, I came to a place where there was a steep down pitch, then a sharp right turn onto a bridge across a narrow canyon. Because of the sharp turn, the bridge could not be seen until almost to it. This time the bridge wasn't there — the county crews had torn it out to build a better one. For the night, they had laid two four by twelve timbers across the gap. By the time I could see the situation, it was too late to stop, so I scooted over on those timbers almost before I realized what had happened. Then I got scared. I made the call, cared for the patient, then told her husband I wanted to wait around a while to see how the medicine worked, though I knew good and well how her asthma would respond, having been through it with her a good many times. He and I sat in the kitchen, our shoes off and our feet in the oven, talking about this and that, until the woman was snoring in her bed, and her husband was nodding in his chair. I could think of nothing but that bridge. At last I was ashamed to stay longer, drove to the bridge, got out, looked at it by the light of the headlights, lined up the car wheels, studied all the angles, trying to get up nerve to go across. You may think a chap that got the wind up about driving his car on a twelve-inch plank, espe-

RESTAURANT

cially when that was a common type of road, should have his driver's license taken away, and you might be right, but I assure you it is different when those planks are 30 feet in the air, and it is a very dark night. It took a long time to get my courage to the sticking point. I had been gone so long that it was getting daylight when I got home, just as my wife was starting to call the sheriff to report me missing.

A wealthy Portland family had a summer home on the beach. The wife and children would come as soon as school was out, and stay all summer. The female members of the family spent a lot of time driving up and down the road in a high priced car; the woman seemed to fancy herself as a speed artist.

The man would come for the weekends, and one weekend, he had an acute appendix, severe. Surgery within a few hours was imperative. He didn't trust country surgeons, wanted to go to Portland. His wife tried to get an ambulance out of Portland but could not. They knew of my makeshift ambulance and asked if I would take him in. She was going to go along, but wanted her car there, so instead of riding with me, decided to drive it. When we got him loaded and ready to start, she ostentatiously said she would let us go ahead; she didn't want to run off and leave us. That sort of riled me, but I made no objection. It was 110 miles to Portland, and more than half of that was under construction, with many, many miles of loose gravel.

That was one trip when the old Hudson strutted its stuff. There was power enough to climb in and out of detours, power and weight enough to plow through that loose gravel like a boat through a pond. It did not take long to lose the wife's headlights. When they didn't reappear, my passenger began to worry and finally asked me to stop and wait. After a few minutes she caught up, and he had some pointed remarks to make about her driving. We repeated that stop three or four times, and each time his remarks to her about her driving became sharper. The last time he really laid it on; I never heard a man talk so to a woman. If I ever talked like that to my wife I'd know better than to come home. Maybe she overlooked it because he was in pain. And she did delay us more than half an hour on that trip. They came back several summers, and I did quite a bit of work for them, but she never mentioned driving again.

Bay City main street, c. 1920, had first concrete paving in north Tillamook Co. Center is Woods Ferry, Big Nestucca River; below, Oceanside road, Seal Rocks, showing plank "paving." (Carroll album, courtesy Tillamook Pioneer Museum)

Tom Wright was a dairy farmer whose house was more than a quarter of a mile off the road, across flat river bottom land. The road back to his house was fair in the summer, but the bottom dropped out when it rained. He solved the problem by running a car from his house to the road, when the ground was soft enough for the car to make a deep track, but not soft enough to stall it. He filled these tracks with gravel, ran over it again to squash the gravel into the mud, then put in more gravel and ran over it. Finally he had a fairly solid road, with ruts four or five inches deep. In these ruts he put an old Chev touring car, and the car would follow the ruts with no help from the driver.

To get to Tom's house in the winter, one would stop by his gate, honk the horn, and someone would come out, crank up the Chev, put it in low gear and head it out to the road. One would meet it before it got to the end of the ruts, hop in, put the motor in reverse, and the car would go back home. Tom Wright's educated Chevy was a real stepsaver, for the only other way to get back there in rainy weather was to walk, and it was hard walking.

Pride of accomplishment is sometimes found in unexpected places. When Highway 101 was built through Tillamook County, there were some rather big cuts and fills. The foreman in charge of one of these jobs was a big, ignorant, hard-driving, hard-drinking, foul-mouthed man. From talking with him, I got the idea that the only things he was interested in were getting out the yardage and the Saturday night spree. He put in long hours and worked hard at his job. One day he told me the amount of his salary and I thought it ridiculously low for a man that could handle such a large crew, so I asked him why he stayed with the job. He studied over that a bit, then said, "I dunno. Guess I'm just a dumb construction bum and don't know nothin' else. But every night at quittin' time, I look over what we have done that day and I think 'I helped to do that' and it makes me feel good."

A Few Cases

In a new community there will naturally be a large proportion of young people, therefore medical practice in such a community will include considerable obstetrics. In the first five or six years, I think I saw nearly every form of non-fatal obstetrical pathology listed in the textbook, probably because most of the patients were having their first baby. However, we got along fairly well, possibly due to a lesson learned from Dr. Boals in those first few weeks. An older man, a Dr. Smith, retired about the time I went to Tillamook County. He had been there 30 or 40 years and had done probably 75% of the obstetrical work of the county in that time, with a phenomenally good record. Dr. Boals told me about this and I asked the reason for that good record. "Damned old fool hasn't got any more sense than to sit around and wait," was the answer. Waiting was always hard for me, but I tried to remember.

The usual obstetrical fee then was $25, but there was one doctor who would do a delivery anywhere in the county for $15. I set my charge at $35 and stuck with it. Some cases were lost to the cheaper man, but not many. At first most deliveries were in the home, with only a few going to Dr. Boals' hospital. In home deliveries, asepsis did not seem to be much of a problem. I tried to observe the techniques learned at school and at Los Angeles County Hospital, but soon saw that the patients got along about the same whether I got there in time to boil up stuff and scrub, or not, in clean homes or dirty ones. Guess the women were used to their own bugs.

Anaesthesia was another matter. We used chloroform for short analgesia, and the husband or a neighbor could usually give that, with me keeping an eye on it, and stopping the anaesthetic every time the patient got beyond the kicking stage. That did not promote easy working conditions, or good aseptic technique, but was reasonably safe. Deeper anaesthesia had me stopped for a time, but soon there came to light a local man, Lem Parker, who would leave his work, or his bed, anytime, to help, so I gave him a little training in anaesthesia. He quickly learned to give a surprisingly good ether anaesthetic. Later, a dentist came to town, and he was a fair anaesthetist. Light for night work always worried me, for frequently kerosene lamps were the only thing available, and ether is very explosive. We never had an explosion, so I guess we worked out the proper compromise between adequate illumination and safe distance between ether can and flame.

One of my first babies was from an Italian woman, wife of a railroad section hand, an easy case. A couple of days after the birth of the baby, as I stopped in to check up, mother and baby were in good condition. Thought I should do something to show I was on the ball, so I asked about feeding the baby.

"I feed him tea," the mother stated. That was not the way I remembered the pediatrics book, so started to object, but she cut me short with "I got eleven bambino. They all healthy bambino. I feed 'em all tea. I know more about it than you do."

She was right—she did know more about it than I did.

One night I was called by a drifting chap to see his wife. The family, including several kids, was camped in an old garage. The case proved to be a tough forceps delivery. The only furniture in the place was an old drygoods box. With the dentist to give ether, we got the woman on the box, and I started to work. It was pouring down rain and the roof leaked badly. I'd give a few tugs on the forceps, then we would drag the box around, trying to get the operative field out of the drip, while I extracted the baby. In three or four days the family moved on.

About 20 years later, an alert looking young fellow came into the office, asked if I remembered the incident, and said he was that baby. His wife was pregnant and he wanted me to care for her. On one visit he said, "Say, Doc, did you ever get paid for delivering me?" On assuring him I had not, he said, "I thought not. The old man never did pay a bill he could run out on."

This boy didn't pay either; moved on a few days after the baby was born. Maybe a dominant trait, following Mendelian Law.

Late one evening there was a knock at the door. When it was opened, there stood a local citizen who said, "Could you come over to my house when you get time? My old woman is awful sick."

To know if I needed to take any special equipment, I asked what seemed to be the matter.

"I don't know, but she seems to think it's a refinement case." It was—her ninth!

The members of one family in the neighborhood were very religious. Their religion was of the emotional type, almost bordering on superstition, placing great emphasis on signs and omens.

A girl of the family, in the middle teens, got herself pregnant. The parents felt the 'disgrace' keenly and made every effort to keep the matter secret. They kept her in the house for the last five months of her pregnancy, even refusing to let her come to my office for the usual prenatal observation, and insisted she be delivered at home. They lived at some distance from neighbors, on the edge of a patch of woods.

She had a long, hard labor, that lasted most of the night. The weather was hot and oppressive. During the night, one of those sudden summer storms came up, the kind that doesn't cool things a bit, but seems to make the heat more unbearable by increasing the humidity. The bright flashes of lightning and deafening peals of thunder mingled with the cries of the poor girl, and the continual prayers of her mother and grandmother. Take it all together, it was a long, trying night, and I guess we were all rather worked up. When the baby finally came, it was dead, had been dead for some time. We wrapped it in a diaper and placed it in an adjoining room while giving necessary attention to the young mother.

Suddenly there was a crash, as of something hitting the side of the house, followed by a scratching noise in the next room, and we all rushed in to see what had happened. We were just in time to see, by a lightning flash, a cougar trying to scramble through the open window, which was six or eight feet from the ground. Evidently the animal had attempted to jump through the window but had only gotten its front feet over the ledge, and was trying to pull itself the rest of the way in. The mother and grandmother screamed, and I think I made some noise, too. Anyhow, the animal backed out, dropped to the ground and disappeared.

I suppose the cougar was attracted by the smell of blood and had decided the little corpse would make a tasty meal. But the women insisted we had seen the devil, come to claim this infant, damned to perdition because of its illegitimate conception.

I always admired those people who could remain calm in emergencies. I never could, though I tried not to show that I was excited or scared. One good example of steady nerves was Amy, who lived in the community for a few years. She had had a little practical nursing experience, and teamed up with a neighbor, Idris, to take care of obstetrical cases in her home. This was early in the town's history. Idris was a worry wart. She could not be still; something had to be done right now; nothing was ever right; she was always in a stew about something. Amy was quiet, philosophical, and never borrowed trouble. We had many hard deliveries in their place. Amy always gave the anaesthetic and Idris did the running around. Many a time, when things were going wrong, and I was getting panicky, I'd look up at the head of the table, and there was Amy, perched on a kitchen stool, chewing her gum, calmly dropping ether, supremely confident that I knew what I was doing, and that everything was going to be all right. Just looking at her quieted me down. That, or something, must have helped, for we did not lose any mothers, and only one or two babies.

One day Amy injured a finger—slammed a car door on it as I remember—and I thought it would be well to have an x-ray. I was going to the hospital that evening and suggested she ride with me. On the way, we made a side trip up in the hills to a logging camp, to see a sick kid. The road was narrow and crooked; it was pouring rain, and in those days automobiles had no windshield swipes. All of this added up to very poor visibility. At one place the road made a sharp turn around a high bank and crossed a canyon on a bridge made by falling a couple of trees across the canyon. A plank deck was laid on the logs and a five or six inch sapling spiked along each side for a guard rail. Because of the poor visibility, I missed that bridge with three wheels. The next thing I knew, we were hanging over the canyon by the left rear wheel. Amy had fallen down under the dashboard, and the only reason I wasn't with her was that I had the steering wheel to hang on to. I was afraid to take a long breath for fear it would jar that wheel off the guard rail.

Going through my mind was the certainty of small town gossip—the young doctor out with another fellow's wife, too busy to watch the road, etc., etc. You know! Amy looked up at me, and as quietly as if she had been in that position many times, said, "I think maybe we'd better get out."

We got out!

A group of men from the camp bunkhouse turned out, got a rope on that Ford roadster, dragged it back on the road, and we went on to Tillamook.

Many families kept a few pigs, feeding them their household refuse, and letting them forage in the woods and the neighbors' garbage. Nobody objected much, for they were good scavengers. Sometimes one would wander down the main street, and if a few loggers were hanging around, it was considered good clean fun to feed him some beer (a half drunk hog *is* a comical animal!). One day a hog had been treated too many times, and was lying in the ditch, dead to the world. A big Swede logger came staggering down the street, saw the pig, and, with the help of a handy hitching post, stood looking at it for a while. Then he said, "Hog," hic, "tomorrow morning," hic, "you bane gonna have the damnedest headache any hog ever had," hic. Then he wandered off down the street.

One Sunday morning, two young fellows who had been celebrating all night, decided to go some place in an old pickup one of them owned. The owner drove, and the other insisted on riding on the running board. After a time he fell off, face down, into the gravel, getting extensive lacerations of his forehead and chin. His companion loaded him in the pickup and brought him to the hospital. For the hour or more it took to clean up and suture the lacerations, he stood in the door of the surgery, as close as we would let him come, bracing himself against the door casing. While I was looking the patient over to see if I had missed anything, and before any bandages were applied, he lurched from the door, took a good look at his friend and said, "How in Hell do you suppose he tucked his nose up out of the way?" That seemed a logical question, with no answer.

One day I was called to check up on the results of a fight. Apparently a lad with too much aboard, got combative, took on more than he could handle, and got thoroughly worked over. He was stretched out on the ground, seemingly unconscious. I started poking around, trying to find any damaged spots and I guess I found one, for the next thing I knew, I was stretched out on the ground, too. That guy, while flat on his back, had literally thrown one from the ground that clipped me on the side of the head. It caught me off balance, and over I went. I did not see it coming.

At first I was mad, but after thinking it over, decided that if I was careless enough not to see one coming, under those circumstances, I deserved what I got.

Sometime in the late thirties, or early forties, one of the Tillamook drugstores changed hands. It had belonged to the same man for many years, and had a large accumulation of out-dated, unsold merchandise in the basement. The new owner started in to clean this out, and found several cases of old-fashioned wine tonics. He knew that nobody ever called for those, but thought he might realize something on them, so piled one show window full, priced ten cents per bottle. Some of those wine tonics had more alcohol than did whiskey, and a lot of fellows about town knew it. Result, quite a run on wine tonics.

The next day, the chief of police came in the store. "I can't make you do it, but I wish you would stop selling that tonic. My jail is full of drunks; I haven't room for any more. And they are all getting their hooch here. You have a perfect right to sell it, but I wish you would stop until I get my jail emptied out and cleaned up."

The Southern Pacific Railroad brought in some Mexican nationals to maintain the track. They were housed in small bunkhouses, made of dismounted boxcars, along the right-of-way. Apparently these boys sometimes got on one another's nerves, for they occasionally stirred up a ruckus. One morning the section boss called me to go with him to check up on a fight in a bunkhouse the previous night. As nearly as I could get the story, all of the boys in this bunkhouse had chipped in to bring a prostitute from Portland to stay with them for a few nights. One fellow had monopolized more of her time than the others thought he was entitled to, so they ganged up to teach him something about fair play. They used knives as teaching aids, and the final upshot was that this fellow had a number of extensive cuts of his hands and arms, but no vital injury. He had gotten in some pretty good licks on the others, too. I thought that at least the one man should be hospitalized, and a couple of the others taken in for repairs. But the boss objected.

"The S. P. don't pay no hospital bills for a fight. Have you got needle and thread to sew 'em up?"

"Yep, I have suture material with me, but no local anaesthetic."

"He didn't have no anaesthetic when he got them cuts; he can get 'em sewed up without it."

So I sat on the edge of a bunk, spread out my stuff on the bunk beside me, and went at it. The young man stood in front of me and held out his hands. It took me over two hours to sew up all his cuts, and in that time he never whimpered or grunted, and his hands didn't shake!

Mucho hombre!

The heavy, outdoor labor seemed to breed a hardy race of men, or eliminated the weaklings. I was not sure just which, but they were a group of stalwarts, who recovered from injuries quickly, and took trouble in stride. I recall one chap who had lost his left hand at the wrist, but was still able to hold a job; could do as much with ax or saw as anybody. He took a contract to get out some shingle bolts, and was working alone. His ax slipped and he cut his left forearm, cutting through one bone and the radial artery. He walked almost half a mile to where he had left his pickup. When he got there, he found he had a flat tire, then discovered someone had stolen his jack and lug wrench. He walked back to his place of work to get his ax, cut a sapling for a lever, lifted the wheel, blocked it up, banged at the lug nuts with his ax until he got them off, changed the tire, drove about 10 miles to my office, then insisted on sitting up while the cut was repaired. Needless to say, he was pretty white by the time the job was finished, but he drove his pickup home and was back at his contract in less than a week.

Another fellow, his arm broken between elbow and shoulder, went home, took a bath and put on his good suit before coming in.

Shingle sawyers work close to a rapidly revolving saw, and most of them ultimately lose fingers. In fact, a few stubby fingers was sort of a status symbol among them. One day a young shingle sawyer was brought in, with the left fourth, third and second fingers severed at the base. I sewed up the stumps and they healed nicely. About two weeks after he returned to work, he was brought in again, with a saw cut about halfway through the first finger. After I got it patched up, and before it was bandaged, he looked it over carefully, then said, "Doc, it's a good thing those fingers are gone, 'cause if they had been there, I'd have cut them off, sure as Hell."

Building the old Neahkahnie Mt. road, 1898. (SP Col., OHS)

The local sawmill built a new smokestack, about 200 feet high. They worked from a circular scaffold, with a platform every seven feet. Before tearing down the scaffold, they sent a man up to paint the stack. While working on the top platform, a gust of wind blew him off, and he fell between stack and scaffold. He grabbed at each platform as he went by, and succeeded in getting an arm over the third or fourth one down. In doing this, he rubbed off quite a bit of hide from the inner aspect of his arm and side of his chest. He pulled himself onto the platform, climbed down the ladder, and came in to have the skinned places dressed. Then he went back to the job, got another bucket of paint, climbed up to the top, and started in where he left off.

Anyone who has driven Highway 101, the road along the Pacific Coast, might well remember the scenic spot where the road goes along Neahkahnie Mountain. Neahkahnie is a spur of the

Coast Range, that juts out into the ocean. The ocean face is an almost perpendicular cliff, about 600 feet high, with deep water at the base. The road plan called for a shelf to be blasted out of the face of the cliff, about 300 feet above the breakers. Two men were drilling holes for the explosives, letting themselves down from the top on bosuns' chairs. One was working a few feet above the other. The lower man looked up just in time to see his buddie topple off his perch. Instinctively he reached out and grabbed the falling man, succeeded in holding him, but dislocated his shoulder in the process. The fellow that had fallen climbed onto the bosun's chair with his partner, hauled both of them to the top, got the injured man into a car and brought him in. Then he went back on the mountain, let himself down over the cliff and resumed drilling, all alone.

Lee Tittle was high up in the hills cruising timber when he developed an acute abscess of the thigh (osteomyelitis). He was too far out to get in easily, so he sharpened up his hunting knife, sterilized it by boiling, and slashed a hole in the abscess. When he finished the cruising job, he came in. He had about a three-inch hole in the thigh.

One day a square-head (Scandinavian) came into the office with blood pouring from a scalp wound. He was a lumber piler, and a sudden gust of wind had blown a board from a nearby pile, hitting him on the head. I stopped the bleeding, washed off enough blood to see what I was doing, and sewed him up. Before he left, I told him he would get two or three days off, but he said, "With a cut like dis on my leg, or my arm, I no can vork, but on my yob a head I don't need. I vork."

These square-heads were pretty tough guys. Without the contribution the Scandinavians made, the timber industry would have bogged down, and maybe the country, too.

Shortly after I opened an office in Garibaldi, a fellow came in with a boil on his neck. He was a stranger to me then, but later I found he lived in town and was a foreman in the mill. He said: "I just want that boil opened. I went to the doctor in the next town and he just picked at it. Guess he was afraid he would hurt me. I'm no baby and I want that thing opened."

That was a challenge. I had a very dull, old scalpel that had been given to me by an older medical friend. He hadn't used it for many years, and I had never gotten around to sharpening it. I took that scalpel and laid the boil wide open. The chap never flinched, paid me and walked out. I felt sort of mean about my unnecessary roughness, but he had asked for it.

Later we became friends; he was a nice guy. Maybe 20 years later we were visiting when he said, "Say, Doc, do you remember that time I came to you with a boil on my neck?"

I remembered.

"You know, I never had anybody hurt me so bad in all my life as you did that day. Guess that's what made me notice you later."

It always surprised me to see how much brute strength some loggers developed. One outstanding example was Big Eric, a Swede, about six feet six, weighing something over 230 pounds, I should guess. One afternoon I was driving up a narrow road gouged out of a steep hillside, when I turned out a little too far to pass another car. The two right wheels went over the bank, leaving the car balanced on the edge. Seemed like a good time to get out. It was near a camp, so I walked to it, getting there just as Big Eric and his gang were coming in from work. They went to the car; Eric climbed over the bank, backed up to the rear wheel, took a good grip (cars had wheels with wooden spokes those days), lifted the car up level and said, "Poosh, boys."

They "pooshed" and he carried one side of that Hudson, which weighed about two tons, until the wheels were back on the road.

A salesman was trying to sell a Chevrolet car to one logger. The man was not much impressed—he thought the car was too light and tinny to stand up. He ended his objection with, "That thing is so light and skimpy built I can tip it over by myself."

The salesman assured him it had a lot of good heavy frame structure that could not be seen. The logger did not bother to reply, just leaned over, picked up one side of the car, and had it almost high enough to fall over when the salesman hurriedly admitted he was wrong.

Dairies and Cheese

Dairying is an exacting business. To be successful, a man must be very regular in his milking, every twelve hours (some milk every eight hours), seven days a week. Most men milk their cows in the same order every time. If a cow is not milked completely and regularly, her ability to produce milk is impaired. A herd of cows is a tougher jailer than the sheriff. There are other disadvantages, too. Cows eat a lot; the more milk they give, the more they eat. Just handling the forage for a large herd is a big chore, to say nothing of paying for it. The investment in buildings, equipment and land is considerable. Good cows are expensive, no matter whether you buy them or raise them. However, there are advantages, too. The dairy industry is usually slower to respond to general financial changes in the country than most industries. People will buy milk in bad times, if they have to leave something else off the grocery list, and they do not radically increase their use of milk in good times. This makes for stability in the industry.

There is an old saying, "If you get a good hold on any old cow's tail and hang on, she will pull you through any kind of trouble." Whether the many regulations of the several supervising governmental authorities help or hinder the industry is a moot question. The answer you get depends on to whom you are talking. Be that as it may, over the years, dairying has been a moderately rewarding way of life for those willing to meet its demands. I have observed that men who stick with dairy cows usually achieve a competence, but few ever get rich.

The original cheese factories were small, located where they were easily accessible to the supporting farmers, and usually located near a stream. They were owned by the farmer patrons, who formed a company, bought stock in it in proportion to their expected milk production. They elected a board of directors to hire the cheesemaker and conduct the business of the factory. After the milk solids were made up into cheese, the residue, whey, was available to the farmers if they wanted it—it was good pig

Dairy country north of Tillamook, top. Dairy herds, below. In earlier days Jerseys and Guernseys were more popular, for they gave milk higher in butter-fat. As popular interest turned away from fat, Holsteins became more popular. (OHS)

feed. Any not taken was run off into the stream. Fishing was fabulous below cheese factories where most of the farmers did not raise hogs.

The farmers were paid for their milk on the basis of the butterfat content. Because of that, much high butterfat milk was produced. Tillamook cheese is made from whole milk and allowed to have all the butterfat that will stay in it. That is one reason why it is such good cheese. After a time, it was realized that considerable fat was getting away in the whey, so separators were introduced, and enough fat extracted from the whey to support a sizable butter industry. Then it was realized that other components of the whey were valuable, and a plant was built to process it, producing milk sugar, or lactose, chicken feed, and I don't know what else.

All of these things tended to centralize the processing of milk. The little factories gradually disappeared, the milk going to a few large factories. The growth of good roads made this possible; it was easier to drive ten miles over a good all-weather road, than it was to drive two miles through a succession of mudholes.

When the Wilson River road was completed, and Portland was only a few hours away, some of Tillamook's milk began to go to the Portland fluid milk market. To participate in Grade A production, much attention had to be paid to sanitary construction and operation. This was expensive but it brought a better price for the milk to those who could afford it and wanted to meet the stricter requirements. A great many of the dairymen thought it was more profitable in the long run to continue producing milk for cheese. This was pasteurized and made into cheese so soon after milking that a higher bacterial count was acceptable.

When I first went to Tillamook County, a large part of the dairy farmers were Swiss. Everybody milked by hand. If milking machines had been invented, they had not reached the Tillamook country. Most people, when milking, grasp the teat as one would a hammer handle, but the Swiss milker doubles his thumbs under. He develops a heavy callus on the outside of the knuckle of his thumb. I never saw such a callus on anyone but a dairyman who milked the Swiss way.

The regularity needed in dairying develops a peculiar attitude. I was much shocked early in my practice to see how milking time

MAPLE LEAF
CREAMERY

had precedence over everything else. One afternoon a man was in the hospital sitting by the bedside of his son, who was dying. After a while he looked at his watch, said, "Milking time," and walked out. The boy died in less than an hour.

Twenty or thirty cows were considered a full-time job for a man, though a few were able to handle more. Gradually the herds increased in size. Milking machines and other machinery increased the efficiency of the workers. Clearing up more land, improving drainage, dyking, better management, increased the amount of forage produced. Strange as it may seem, in a land of such heavy rainfall, irrigation proved very profitable. The productivity of the cows increased as more attention was given to breeding practices. All these factors boosted the milk flow.

One family I knew had four husky boys, high school age or older. With the help of these boys, the father cared for a string of something over 120 cows. The boys all came down with the mumps the same day. The father, with the help of a high school girl, milked all those cows for three days, until they could get help. When I saw him, his hands and forearms were swelled as large as hams.

The importance of regular milking is understood by all dairy families. One afternoon, about milking time, I stopped at the farm of a friend, but there was nobody home. Found out later that they had gone to town and their car had broken down, making them late. As I was getting in the car to drive on, their two grade school boys got off the school bus. They looked around, saw milking had not started, changed their clothes, got the cows in from the pasture, put them in the stanchions, and started in to milk that string of over 50 cows. They had a good start on the job, too, when their parents and older brothers came home.

Dairy cows don't look very tough, but I guess they are. One chap had a farm upstream from a high fill carrying a road across the little valley. One night a storm and high water brought a lot of debris down the stream. Some of this debris caught in the bridge, plugging it and damming up the stream until the water was high enough to run over the fill, flooding several square miles, much of it quite deep. Because it was a stormy night, the farmer had left his cows in the barn. When he discovered what had happened, his cows were standing in water half way up their

Cheese factory, c. 1915. Note size of stumps along road. (OHS)

sides. Since his buildings were on a little knoll, he estimated that the water was at least ten feet deep for the half mile or more to the nearest hills. He climbed up in the hay mow, and tore up the floor over the cows' heads. Then he used sections of his hay rope to lasso each cow, pulled her up until she had her fore-feet in the manger, then tied her up so short she could not get out of the manger. At milking time, he waded in, reached down, and milked under water. There was no possible way to feed the cows, and they had to stand in an unnatural position. They were there three days and he lost one, from pneumonia. They were back to their usual production level in about a week. A neighbor tried to swim his cows out, and over half of his herd drowned.

One of the hazards of dairying is caring for the bull. Many bulls of the dairy breeds are vicious, and all are potentially dangerous. Every year a few men were injured or killed by their bulls, frequently one thought to be gentle. The universal weapon against bulls was the pitchfork, and many careful farmers made a practice of carrying one in their hand every time they had to go near "The Mister." Any bull that had had an experience with the business end of a pitchfork, had a healthy respect for it.

Since the cows brought in the money, they got the choice care and feed. The bull got the hay the cows left in the manger. The grass along the road was too dusty for the cows, but was good enough for the bull, so he was tethered to a fence post with a chain long enough to give him some freedom of movement, but short enough to keep him out of the traffic. On asking directions, a stranger was likely to be told, "Sam Smith lives nine bulls down the road." Of late years, the practice of artificial insemination has become common, so it is unusual to see a bull alongside the road.

One young fellow I visited frequently was just getting started dairying. He had a small herd of better than average cows and wanted a better bull than the one he had. He could not afford to buy the quality of animal he wanted, so he bought a bull calf with the desired blood lines. This calf was raised like a pet, with the run of the farm. Really, he was treated more like a good dog than a calf. I visited the farm when the calf was about eight months old, and noticed he was confined and that his owner picked up a pitchfork before entering the pen. I asked how come? "Oh, he's just beginning to find out he's a bull," was the answer.

Tom Elkins lived in a house that was some distance off the road, with lots of trees around it. He bought gas for his truck, tractor and car in 50-gallon drums. He noticed a drum of gas didn't last as long as it used to, so began to keep a record and found he was using only 30 or 40 gallons from each drum. He thought about watching his gas barrel but liked his sleep too well. He had a really mean bull, one that would charge anyone who came near. He never approached it without a pitchfork in his hand. So he tied this bull to a tree by the gas drum. About two o'clock, there was a terrible commotion out in the barn lot. By the time he got his pants on and a lantern lighted to investigate, there was nobody around, but the bull was pulling on his tether, pawing the dirt and bellowing. Next morning there were two five-gallon gas cans a short distance away and a piece of cloth, that looked like it might have been torn from a shirt, caught on a barbed wire fence.

Nobody ever claimed the gas cans, and he missed no more gas.

Howard Tilden, a logger, told me this one. "I worked for Markham and Callow when they were in here. I was on the log dump, so had to work with the tides, and sometimes we would be late getting through. There was a pasture between camp and the log pond, but we had permission to cross it. The owner of the pasture had a Jersey bull he sometimes put out there at night, and he was a mean one. We were afraid of that bull. During the daytime, there was no problem, but night was something else. We never knew when that bull might be there, so the company gave us permission to drive a company truck when we worked nights. We figured we would be safe in the truck.

"The man that drove the truck was a young fellow who knew nothing about bulls. He was skeptical of the danger. At first the bull paid no attention to the truck. This went on for a few nights, and the driver kept talking about how that bull was a coward. One night, as we were going to camp, we saw him about a hundred yards from our road.

" 'I'll just give him a scare,' said the driver, and before any of us could protest, he turned off the track directly towards the bull with the horn going full blast. That was too much for Mr. Bull. His head went down and his tail came up. With a bellow he came charging at the truck and met it head on. The truck's radiator was ripped off, fan and fanbelt ruined, headlights smashed, and a few other things damaged. The bull was pretty well damaged, too. The boss was sore, not only at the repair bill on the truck, but the fact that it was out of service for a couple of weeks. And he didn't cool off any when the farmer came with a bill for the bull. He was so badly injured he had to be killed. The tracks, broken glass, and other debris made it pretty clear what had happened, so there was no question who was responsible. He threatened to cut off our right-of-way, so the boss had to pay, but he wasn't happy. He fired the driver and was going to fire the rest of us, but we finally convinced him it was done over our protest."

Not all bulls were vicious. Vern Larson told this tale:

"When I was a boy, my father had a big Holstein bull that was a pet. We kids rode him and played with him like a pony. It was my job to stake him out along the road every morning.

"The schoolhouse was just across the road from the cheese factory, so we always tried to be ready for school when Dad took the milk to the factory. If we weren't ready when he took off, we had to walk, and it was about two miles.

"One morning I was late, overslept or something, I guess. When Dad was ready to start to the factory, I was all ready but hadn't staked out the bull. It was quite a piece to the stakeout place for that day, and I knew I couldn't lead that pokey old bull that far, tie him up, and catch Dad's wagon, so decided to lead him behind the wagon. Maybe Dad would drive slow enough so I could run and catch him after I'd tied the bull's rope to a fence post.

"We started off, and something displeased that bull. Maybe the horses walked faster than he wanted to go, or maybe he just didn't like to be led behind a wagon. Anyhow, he stepped forward, put his horns under the rear of the wagon and lifted. Over went the wagon, cans of milk, boys, Dad, all in a pile on the side of the road. Then the bull started cropping grass. He wasn't mad, he just didn't like things as they were and did something about it."

One farm practice that was new to me, one brought over from Switzerland I understand, was the use of liquid tanks. These were tanks of wood or concrete, so placed as to catch the drainage from the stables, the urine, and the water used to hose down the floor. Some of the solid droppings were washed in, too. These tanks were large enough to hold the accumulation of several weeks. This liquid would undergo a fermentation process, in the course of which a crust would form over the top, strong enough to hold up a dog, unless he happened to step on a weak spot. After one such experience, Mr. Dog could not be induced to walk on that crust again. This liquid is not offensive as long as the crust is not broken, but is exceedingly so if disturbed. The liquid is pumped into a tank mounted on a wagon (later a truck), which has a distributing device on the rear end, and scattered over the fields. It is wonderful fertilizer, really makes the grass grow, but you can smell it for a mile, until the next rain.

One of the first liquid tanks in the area was built by John Naeglie, a Swiss dairyman who lived a short distance east of Tillamook. He built the tank of three by twelve planks and caulked the joints with moss to make it water tight. It requires a special pump to handle this liquid. At that time, none were made in this country, so he sent back to his native land for one. It was shipped from the factory to a European port by rail, by boat to New York, by rail to San Francisco, by boat to Portland and by truck to Tillamook. The freight charge from Portland to Tillamook was greater than all the other freight charges put together. That is a good example of how the lack of good roads can isolate a community.

Seeing how the tank liquid made the grass grow in the pastures, I got the idea it would be good for my lawn. That was a mistake. I had thought I could control the odor by heavy sprinkling, washing the odorous material into the ground. I could, but it took about a week to do it, and in the meantime I was decidedly unpopular with both my family and the neighbors.

The other part of the mistake showed up soon. I had to mow that lawn every other day all summer. I took off enough grass to have fed a cow. And that was before the day of the power mower!

One dairy rancher I knew had a good place on a river near enough the mouth that salt water came up every high tide. If it had not been dyked, the farm would have been worthless. A slough from the river cut his place in two. The water level in this slough was controlled by a tide gate. The slough was shallow enough that the cows could wade it, and a rickety plank footbridge took care of people. Carp had come into this slough from the river, and it was a good place for them. They are vegetarians and they would root around in the mud bottom of that slough like a bunch of hogs. There was a lot of feed for them, and many were as large as 18 or 20 pounds. The farmer didn't like these fish; they kept the water so muddied up the cows would not drink it.

He had a dog, a black German shepherd, that liked to catch carp. This dog would wait on that footbridge, near the end over shallow water, and when a carp came along would jump in. Carp have large, thick scales, and a plump, well rounded body that tapers down just ahead of the tail. He seldom missed getting that tail hold when he dived in. Then it was a pulling match, the dog trying to pull the fish out on the bank, and the fish trying to pull the dog into deep water where he would have to release his hold or drown. It was an interesting battle, and the dog usually won it. But not always. Some of those fish were too big for him. I saw some that must have weighed over 30 pounds.

When he got one on the bank, he would watch it flop. If it seemed to be making progress toward the water, he would grab it and drag it back. I never saw him eat any of a fish, apparently he didn't like fish, he just liked to catch them. While he was watching a fish die, he would strut around, comically proud of his victory, and was always eager to try it again. But when he tackled one that was too much for him, he would crawl out of the slough, and without even waiting to shake the water out his hair, would tuck his tail between his legs and head for the barn, where he had a hiding place under the floor. No amount of coaxing would induce him to try again for several hours. He could be ashamed as well as proud.

This dog did something else I have never seen another dog do. The owner had about 40 cows, all of them named. He could tell Laddie to go out in the field and get Lottie, or Peggy, or Brindle, or Daisy, or any other and pretty soon the named cow would come up, the dog nipping at her heels if she showed any tendency to loiter. I've seen a number of dogs that would bring up a cow or sheep when the desired animal could be pointed out, but this dog would get a named animal that was completely out of sight. There are probably others, but I never saw one.

One day Lige Sholberg was in the office. Lige was a likeable young fellow, with a quiet sense of humor. I wasn't busy, and he seemed in no hurry, so after I answered the question that brought him in, we sat and talked a bit. He had a farm on the lower Tillamook River, and he got to telling me about his neighbor who lived on the adjoining farm.

"Bill is a suspicious fellow. He thinks everybody is trying to work an angle. He has not been farming very long; born and raised in town, and worked in a store until he moved on that farm. Don't know how he got it; don't think he had money enough to buy it, but he managed some way. He didn't know much about dairying, but he learned. Really did pretty well. There was a good herd on the place when he took it over, and he took good care of it. He did a lot of funny things but I guess that is part of learning. But he never got over his jealous nature. He had more cows than I did, but I regularly put more milk into the factory, and my test nearly always topped his by just a little bit; he had Jerseys, too, and I didn't. That made him mad. If one of my cows got through the fence into his pasture, he got mad, but if one of his got in my pasture, it was a good joke. His dog and my dog would fight every time they got together. They were about the same size and pretty evenly matched. If they fought at his place, he would interfere with a club, rescue his dog and chase mine home. When they fought at my place, I'd just let them fight; figured that the sooner they settled who was boss, the better. If Bill heard the racket, he'd come charging over to break it up, and was mad because I hadn't done it. That kind of a guy.

"There were a lot of muskrats along the river which had their burrows in the dyke. I was trapping them, making quite a bit of extra money. Lot of extra work, too. When Bill first came on the place, he told me he would shoot anybody that set traps on his land. I remembered that, but he had a lot of rats in his dyke, and I wanted them, so I'd slip over at night and set traps, and run them after dark. I had a lot of traps on his land but he never found any. Of course, I didn't trap his land as hard as I did my own. I actually had traps under his barn. Caught three muskrats, a skunk and a coon there, and he never suspected.

"One day he came over and told me he had noticed that something was digging a lot of holes in his dyke, but he noticed that mine was not so badly damaged as his. I told him yes, I knew it. It was muskrats, and I was trapping them. The skins weren't worth much but they paid for the labor, and it protected my dyke. (It was war time and the price on skins was the best I had ever

Interior Maple Leaf Cheese factory, Tillamook, c. 1915. (SP, OHS)

known; I was making almost half as much off those rats as I was off my cows, and I didn't want him to know that.) Then he asked if I would come over on his place and trap his rats. When I told him I had trapped some on his place he got mad and told me to stay off his place. Pretty soon he cooled off, though, and repeated his request. So from then on I could run my traps on his place in the daytime. A few months later he found out what I was getting for rat skins, and he was mad. Told me to stay off his place; he would trap his own rats. From the looks of his dyke, don't think he caught many.

"That kind of a guy.

"He sent away and got some fancy Indian Runner baby ducks. He raised them in a pen in his back yard, and when they were big enough to look out for themselves, turned them loose to rustle their living. There was a lot of stuff around for them to eat. They were beautiful ducks.

"Next spring when they started laying, some of them stole nests on my place. I didn't say anything, just gathered the eggs, and when I had enough, I'd set them under a hen. I had five hens setting on duck eggs before those ducks quit laying. When they hatched out and were running around the place, Bill kept looking at them and finally asked me how come I had Indian Runner ducks too. With a perfectly straight face I told him I liked the looks of his ducks so much I got some eggs and hatched them under hens. Didn't think it was necessary to tell him where I got them; he would get mad. It made him feel good that I wanted something he had and he was real friendly for a while.

"A little before Thanksgiving, we both sold our ducks to the same butcher, and mine outweighed his a little. Guess I'd had a little more experience in feeding than he had. But he was mad.

"That kind of a guy."

In the Tillamook area, where most of the agricultural income is from cheese, everybody realized that the only way they could compete against cheese made closer to the centers of population, was to make a better cheese. Tillamook farmers took a great deal of pride in their cheese, and in the fact that their product regularly ranked high in the national contests. (It really was, and is, fine cheese.) They went to a great deal of trouble to produce milk that was of low bacterial count and free of odors that might affect the cheese.

The intense rivalry between the various factories came to a focus at the county fair. The cheesemaker from each factory presented a sample of his product, a "wheel," weighing about 30 pounds. These samples were placed in a row on a table, and covered with paper. Each cheese was given a number, but only the fair official handling the contest knew which number marked whose cheese. Then each cheesemaker, acting independently, took a sample of each cheese by sticking a sampling instrument through the paper, and rated the sample on a percentage scale. The total of all scores for each cheese determined the final rating. The scoring was conducted in public and attracted quite a gallery. The scoring record was published in detail in the local press.

One year, everybody got a good laugh because the chap who had been champion for several years was not anywhere near the top. The detailed report of the scoring revealed it was his own report that put him out of the running. Nearly all the others gave his cheese a good score, but he just happened to get a little sample that was a bit off and he scored it down.

Minor Products

Tillamook County has no mineral industries of value. There is a deposit of low grade coal in the north end of the county, on Coal Creek, that was worked sporadically, but I do not think it has been touched for many years. There are extensive deposits of clay that might be suitable for pottery. There is sand and gravel for local needs. There is a great deal of rock, but it is of poor quality. Rock for the recent jetty construction had to be barged in. A very small deposit of mercury was said to have been found during the second World War.

Among the legends is one that in the early days of the settlement, the Indians brought in, to trade, sizable nuggets of gold. They would never tell where they got it, except that it was somewhere on the divide between the Willamette Valley and the coast. That is a big territory, and rugged, but many men looked it over. There is a story that one man befriended an Indian woman who was old and sick. Just before she died, to show her gratitude, she told him of this place where the Indians obtained gold. "Where water runs into a lake in a black canyon." This man spent many years searching but never found it. As far as I know, no one ever found that old Indian mine, if there ever was such a thing.

The gold fever took a sudden spurt in the early twenties when the story got out of an old fellow found dead in a Tillamook boarding house, and concealed in his room was a quantity of gold dust and nuggets. Maybe so, I don't know.

I do know that around the mouths of some streams, where they flow into the ocean, there are deposits of black sand that bear some gold, but it is so finely divided that it cannot be recovered by usual gold mining techniques. Many have tried but I never heard of anyone recovering enough gold to make it profitable.

Along with the story of the lost gold mine, is the story of the buried treasure on Neahkahnie Mountain. Early settlers were told by the Indians that their ancestors had seen a big canoe with wings come near land; a small canoe came from it and white men, one

Neahkahnie Mountain, 1909. (Gifford, OHS) Lower right, inscribed rock, long considered clue to pirates' buried treasure. No one has solved problem yet.

with black hair and beard, the other with red hair and beard, and two black men came ashore. The black men carried a heavy box up on the mountain side, dug a hole and buried it. Then the white men killed the black men, went back to the big canoe, and sailed away.

Ever since white men have lived in the Tillamook country, efforts have been made to find this buried chest. The assumption is that these white men were pirates burying loot. Every once in a while, an old map turns up, supposed to be authentic, and somebody gets excited. The area is dotted with old diggings; thousands of dollars have been spent, and at least two lives lost. Such items as Chinese coins, bronze chest handles, scraps of rusty iron, have turned up. Rocks with marks of unknown meaning have been found at various places on the mountain side. Interest still persists; within the past few years, I have seen an article in a national magazine and a national television program about the buried treasure on Neahkahnie Mountain.

Better authenticated is a story told by the Indians of a ship wrecked near the same location, and white men being washed ashore, along with quantities of some material strange to the Indians. Some of these men survived, were cared for by the Indians, later left to the north, and were never heard of again. Supporting this story are reports of a few red-headed individuals among the normally black-haired Indians; supporting the story of the new material—over the years pieces of beeswax have been washed ashore. Some of these pieces are quite large and bear imprints of religious emblems. Analysis shows that the material really is beeswax that came from southwestern Europe. This is proved by inclusions of pollen, plant pieces, and insects found only in that area. There is history in Spain of a ship loaded with supplies for the Catholic missions on the west coast of America, which disappeared. Maybe this was its end. Anyhow, after every storm, people patrol the beach, hoping to find another piece of beeswax. Sometimes they are successful.

Beachcombing was a popular sport in Tillamook County as well as on all other beaches. There is no telling what a storm might cast up. One fellow I knew found a large pile of sawed lumber, evidently part of a deck load that broke loose from a coastwise ship or raft. He salvaged enough to build a house. Saw logs have been found, as well as Japanese net floats, teakwood, pieces of ship equipment, shells, and all sorts of flotsam and jetsam. Sometimes a large marine animal was washed ashore, and disposing of it was a problem.

While the principal products of Tillamook County were timber, cheese and fish, there were a number of other things, some interesting and unusual.

Chittim bark, *cascara sagrada,* is the source of cascara, a laxative still widely used. It is a shrub, or scrubby little tree. In the spring and summer, the bark peels off fairly easily in large pieces. These are spread out in the sunshine to dry, then sacked. If he could find a good patch of chittim, an energetic person could do well. Most of the peeling was done by women and children, making a little spending money, but a few men did it full time during the season. The big problem was not to peel the bark, but to get it out. Most of the good stuff was pretty far up in the hills, and most of the trails were too rough for pack horses.

I talked to one peeler, who was quite proud of the way he had solved his transportation problem. He had gotten hold of a burro to pack out his stuff. I asked how much he could bring out at a trip, and with a big grin he said, "Me and this here other jack, we can bring out 400 pounds to a trip."

Just before the county fair, lots of farm children got out and gathered chittim near home, for money to spend at the fair. The children of one family I knew, one year had gathered quite a pile. When this bark dries, it curls, and takes up a lot of room in the sack. To save space, it is broken up into small pieces. A corn sheller does a pretty good job of breaking up the dry bark. On the day before the fair, these kids were putting their bark through a sheller, when a little girl, about eight years old, got her right hand caught. Before she could get it out, she lost all four fingers and the thumb at, or above, the first knuckles. She had just started taking piano lessons, and her major concern was the effect of her accident on her piano playing. I wasn't able to encourage her much, for it didn't seem possible that those little stubs could ever learn to play a piano.

Five or six years later, I was making a call on a neighbor, and when I came out of the house, this little girl, now a big girl, was waiting by my car. She asked if I would like to hear her play the piano. Of course I would. She played several pieces for me, none very complicated, but she did them acceptably. I've heard better piano playing, but I've never heard a concert I enjoyed more. Learning to play the piano is a big job when you have two good hands, but when you just have little stubs on one hand, it is an achievement. I have not seen the girl since that concert, but I often think of her.

Another medicinal produced in Tillamook County was foxglove, the source of digitalis, a heart medicine. Foxglove comes up each year. There is a central stalk that sometimes gets to be six or eight feet high, but is usually three to five feet. Around the stalk are large, thick fleshy leaves, and there is an attractive bluish flower at the top of the spike. If not checked, it is a pest; it will crowd out everything else. Before there was a local market for it, the farmers put up a losing battle to keep it out of their hill pastures.

At Bar View there was a retired sea captain. He had bought a chunk of hill land and raised sheep and beef cattle. He was a nice old fellow, but his speech was a bit salty. I saw him frequently for a chronic heart condition, and prescribed digitalis. He finally became aware what it was; maybe I told him. When I'd come into his room he would hold up his bottle of pills and say: "I've been fighting this —— stuff in my fields for the last ten years and now I have to buy the ——— stuff at a ——— drug store."

Ferns grow profusely anywhere in the woods where a little light can get in. The florists use the fronds in their arrangements so some years there is a good market. The pickers snap off the fronds near the ground, tie them up in the packets, and make them up into packs to carry out.

Sometimes there was a market for huckleberry and salal foliage to be used by the florists, but the demand seemed to be spotty.

Blackberries grow abundantly in Tillamook County. There are two kinds: one is a real wild berry, a small fruit, very tasty, growing on a small bush. These are found only away from civilization, in open spaces in the hills, such as old burns. The other is a domestic strain, the Evergreen, that has escaped and grown wild. It is found along fences or any place it can get a foothold. If it gets into a field and is let alone for a time, it will take over. The better the

Picking cranberries at Sand Lake, Tillamook Co., 1912. (OHS)

soil, the more vigorously the Evergreen grows. The canes are large, sometimes 15 or 20 feet long, and have heavy, sharp stickers. The berries are borne profusely and are of good size. They are much in demand by the canners, pie makers, jam makers, and I think some go to the distillers. One year I heard they brought $180,000 into the county.

For a time, daffodils and lilies were raised commercially, but as far as I know, not lately.

Around Sand Lake, they raised some cranberries. They were of good quality, and found a ready market through the national cooperative organization. I found that the last of the crop to be harvested had the best flavor. Now that they are picked by machine, I presume they are all picked at one time.

The mild, damp climate is favorable for the growth of holly, and a fine quality is produced. One county agent was enthusiastic in his promotion of holly and I tried it. What he did not tell me—maybe he didn't know—is that deer are crazy for holly. Or, maybe I was the first to plant it close to the timber. In spite of the fact that there was a loud-mouthed dog on the place, they would jump the fence, grab a few mouthfuls while the dog was getting to them, then sail over the fence and into the brush. They killed the first two rows of trees next to the woods, and severely damaged others.

Some little trees, two or three inches thick, they ate back to a stub.

As elsewhere, some Tillamookers tried raising fur. The principal variety was mink, and there were a few sizable operations. Fox, fitch and chinchilla were tried but as far as I know, nobody stayed with them very long.

The first cattle in Tillamook County were beef and draft animals. The large, heavy animals made a lot of meat or made big oxen. While later the dairy animals became much more numerous, there were always some beef cattle ranging land too rough for dairy cattle. In later years, since dairying is not quite so attractive, more and more land is being turned to the beef cattle.

Since Tillamook had such a good reputation for dairy cattle, people wanted some of their stock, and a lively business in little calves was built up. Most of the dairymen raised a few of their heifer calves, those from their best cows, for replacements in their own herds. It was a lot of bother to care for calves, and the milk they drank was salable, so most of the calves were surplus. They were killed soon after birth or sold for a pittance to the mink men. Now there is a market for all the heifer calves. They go to central and eastern Oregon and Washington, possibly elsewhere.

When the cheese factories had no market for the whey, quite a number of hogs were raised, but when that source of feed was cut off, the hog business died. There is no feed grown in the area suitable for hogs.

All grain fed the dairy cows is shipped in, for grains do poorly. Some hay is made, but many years it is difficult to get it dry. When the farmers found out how to put grass in a silo without drying, it solved one of their big problems. In spite of that, much forage is shipped in, hay in bales and cannery refuse, mostly corn, to go in silos. The principal agricultural product of Tillamook is grass, and the cows harvest that.

An interesting, but very minor, product is evergreen tree cones. Government agencies and nursery owners buy them for the seed. My impression was that most of these were gathered by women and youngsters, to make a little extra money.

Forestry

Much of the land of Tillamook County is not, and never will be, fit for anything but growing trees. It is too rough even for grazing. As that type of land is logged off, efforts are made to reforest. There are several ways to do this, the most common being to leave enough trees to produce seed. These are left on ridges where the wind and water will help scatter the seed. When there are no seed trees available, as after a fire, seeds gathered elsewhere is scattered over the ground. Formerly this was done by hand; now it is done by airplane. Also, little seedlings, raised in nurseries, are hand planted. This is a good way to handle small areas, but is not practical for large areas—too expensive.

Many people are critical of the wasteful logging practices, and no one can deny there has been much to justify such criticism, but there is something to be said for the logger, too. In the past, the American public would use only the best grade of lumber. There was no use bringing in logs that would not yield good lumber. Things are changing in that respect; poorer grades are being used for many purposes, and pulp products and compressed board leave no excuse for excessive waste in the woods.

Timber is a crop, just like corn or wheat. It has a time of growth, a period of maturity when it is ready for harvest, and a period of deterioration, leading to ultimate decay. True, timber may take a hundred years to go through the cycle corn goes through in a few months. But when it reaches its time of maturity, it is ready for harvest, and if not harvested will deteriorate. Much of the virgin timber, at least in Tillamook County, was ripe or overripe, and should have been cut many years before.

Selective logging is advocated as a solution to the problem of perpetual supply of timber, and is being practiced by the government and some large companies who are managing their holdings to insure a perpetual supply of logs.

In areas so steep the only way to get logs out is by cable equipment, complete logging, called block logging, leaving seed trees, is the only way to do it. On such rough ground, any other type of logging would knock down, or break, too many small trees. I have heard that helicopters are being tried for logging; if they are successful, selective logging of steep ground may become practicable.

Shortly after I landed in Tillamook County, I heard an experienced timberman estimate that only ten or fifteen percent of the timber on the hill ended up as boards in the mill yard. Knotty,

crooked trees, tops, broken pieces, dead timber, stumps, etc., could not be used then. Advances in technology are changing that. Better logging methods, better utilization of poor material make more and more timber usable. One has only to watch truck loads of logs go along the road these days to realize what a tremendous quantity is now being used that would have been left on the ground a few years ago.

In areas where the terrain is suitable for agriculture or grazing, the land is not reforested, but is cleared. Naturally, the best land is cleared up first. When the logger has finished, an incredible amount of debris is left on the ground, but grass will grow around it, and often it is left that way, just waiting for it to rot. And that takes a long time, but is the only practicable procedure for marginal land.

I remember asking a Swedish dairyman, on whom I was making a house call, how long it would take for several large stumps in the field across the road to rot away. "I don' know, long time I think. I bane here thirty year and they look yust the same to me."

If the farmer just wanted to tidy up a bit, he would throw the pieces he could handle against the stumps and logs and burn them, hoping to whittle down the big ones a little. If he intended to do a good job of clearing, he left the small stuff for later use. In this case, he would cut the chunks of wood into pieces he could handle with whatever power he had available. The stumps would be torn to pieces with dynamite, then handled like the other chunks. Originally, the only power available, aside from the farmer's back, was oxen, then horses, then small steam donkeys, then large donkeys, and last, the bulldozer.

When plenty of power was available, a tree would be rigged up like a spar tree in a logging camp, and all the debris within reach of the line, pulled up into one pile to make a really big fire, when it was safe. I have seen these chunk piles 40 or 50 feet high and covering half an acre or more. Clearing land this way was expensive, but the results were good. The rich bottom land was valuable enough to justify such an expenditure.

If the land was near the coast, it was sometimes necessary to dyke it. It is surprising how far up the stream one will find salt water on the extremely high tides. Much of the best land is low enough that a high tide will flood it. If, once or twice a year, a high tide backed salt water over a field, no good grass would grow there. Dyking called for building a tide gate where the dyke crossed a stream. A controlled exit would be built in the dyke for the stream to flow through; if small it would resemble an ordinary road culvert; if larger, a timber or masonry passageway. From the top would be hinged a solid door or gate, swinging outward. The force of the current would keep this gate open so long as there was no pressure outside, but if the tide came up to the gate, it would close from the outside water pressure. Of course, the stream flow was cut off, and water backed up behind the gate. If fresh water backed up over the land, no harm was done. When the height of the fresh water was great enough, it would force the gate open, but the outward rush of the fresh water under pressure would keep the salt water from coming in, until the pressure was equalized, when the gate would close again. It was a simple arrangement, but it worked. Its only hazard was that a piece of driftwood might prevent the gate from closing. That had to be watched.

Originally all this had to be done by the landowner, and the amount of labor and expense sometimes put it beyond his ability. In later years, the government, through the Soil Conservation Service and other agencies, has actively promoted land improvement, with advice, engineering help, and loans to buy supplies, or to hire heavy equipment.

Tillamook Co. land clearing; roots hauled up in 100-ft. pile with donkey engine and pile set afire. Note two men.

Fishing

People who have not seen it can have no conception of an old time salmon run. The sheer number of fish was unbelievable—there just didn't seem to be room in the water for all of them. Oldtimers have told me that at the height of a run, it was impossible to use a boat; one couldn't handle the oars. They also told me it was possible to cross an unfordable stream by stepping quickly from one fish to another. Of course you got wet feet! These statements are hearsay. I can accept the first; the second seems a bit improbable, but the fellow who told me was a deacon in the church, so I suppose it must be true.

Be that as it may, even as late as 1923, the runs were tremendous. The fish came into the streams in untold numbers, purposefully heading upstream, like traffic on a busy one-way street. Each fish is headed for a certain tributary, and when it reaches it, turns off, like the evening traffic into the suburbs of a city. They travel steadily until the water begins to get shallow, then they rest in the deeper pools. When rested, they will make a run at the riffles, forcing their way through water so shallow more of their bodies are in air than submerged. They will force their way up little brooks that look hardly big enough for minnows. They will go over dams several feet high, getting up speed in the deep water, then swimming right up the falling water. If unsuccessful at first, they will try again and again.

Though a few individuals of some species spawn and return to the ocean, most salmon come up the streams to spawn and die. Along some streams, literally tons of decaying fish lined the banks after the runs, creating a stench almost unbearable until the seagulls cleaned up the mess. If the run coincided with a freshet, the height of the water was marked by dead fish caught on the barbed wire fences.

A run is still an impressive sight, but civilization has taken its toll. New regulations in Oregon (and probably the other Pacific states) should allow the depleted runs to build up. Until some way can be found to control stream pollution and off-shore fishing, the runs can never approach those of earlier days. Maybe it is not desirable to have such heavy runs.

In one place, Highway 101 crossed the lower part of the valley of a little river, the Kilchis, on a grade about a quarter mile long. The road has since been relocated. The bridge over the river was not adequate to take the flood run-off, so the land upstream from the grade would flood until water ran over the grade, sometimes several inches deep, but seldom deep enough to stop traffic that really wanted to get through. As long as you could see all the fence posts along the grade, you could get through. If some of the fence posts were under water, better not try it. If a flood of the Kilchis happened to coincide with a salmon run, the fish would go over that grade like going up a waterfall. If one of those big fish, weighing up to 30 or 40 pounds, going at their usual speed, happened to hit the front wheel of a car when the driver was not prepared for the shock, it would jerk the steering wheel right out of his hands, and off the grade he would probably go. Once when I waded out on this grade to help get a car back on the road, a fish hit my legs and knocked me down. I have been told that at times it was necessary for the State Police to put a net below that grade to stop the fish. After an interval, they would stop the highway traffic and let the fish through. Sounds like a tall tale, and I didn't see it, but I have talked to reliable witnesses who vouch for it.

A friend of mine once found a large salmon wedged between the hood and front fender of his car after a trip through this flood.

A construction man told me he once worked on the first railroad bridge across the Kilchis River. A freshet brought a lot of brush and other drift against their scaffolding. This let the water through but the interstices were not big enough for the salmon to get through. Before they got the stream cleared, a salmon run started. The fish accumulated below the obstruction until there was not room for them in the water, and they were pushed up on the gravel bank by the force behind them. This continued until they could bring up some heavy equipment to drag that brush out of the water. Literally hundreds of dead fish were left on the bank. These had to be gathered and buried, or the stench would have stopped the job.

Typical bay fishing boat and net. (OHS)

When I went to Tillamook County, I had no conception of the size of chinook salmon. One of my first patients was a fisherman with an infection of the hand. When we were finished, he told me he had no money, but if I liked fish, he would give me some on his account. That was O.K.—I like fish. Next morning, I drove to his house and he brought out a salmon weighing almost 40 pounds. I protested that was too much fish but he said, "That was the smallest I caught last night."

Was I impressed! Found out later it was the only one he caught that night. The run was over. This was just a stray that had gotten behind the others.

Anyhow, it was a lot of fish for two adults and a year old baby. That was when we learned how to can fish. And it kept, too.

Fishing in Tillamook Bay was never as big an industry as it was along the Columbia River or in Alaska, but at one time a great many people made a living at it. There were canneries at Bay City and elsewhere that processed the catch, but as the runs dwindled, they closed and the fish were shipped under ice, mostly to Astoria, I think.

There were four distinct varieties of salmon that came into Tillamook waters, and they came at different times of the year, though there was sometimes a little overlap. The chinook was the largest variety. I have seen specimens weighing over 50 pounds, but the average was smaller. The silverside, or coho, was smaller, and the chum, or dog salmon, was still smaller, averaging around 10 pounds. The steelhead was more a sport fish, but I think the cannery accepted them if they were caught with the other varieties.

Nets for salmon fishing were drift nets and set nets. The drift nets were laid out in the main channel of the bay and allowed to drift up or down with the tide. Set nets were stationary; one end was fastened to some object, as a tree or stake, and the other laid out into the bay or river. If in a river, the law limited the length of net so the stream would not be completely obstructed.

In each case, the nets were a network of heavy cords, cotton or linen, the length and width depending on where used. There were ropes along top and bottom. On the top rope, were threaded wood floats; on the bottom, lead weights. These floats and weights caused the net to float upright in the water.

At first these nets were woven by hand and weaving them was a between season job. Later, machine woven nets were introduced. There were many arguments as to the relative value of the two kinds, but my conclusion was that the choice was influenced more by other things, than by the relative fish catching ability of the two kinds of nets; things like, did the fisherman have money enough to buy a ready-made net, or whether he was industrious enough to weave his own.

The size of the mesh varied with the variety of salmon running. If the mesh was too small, the fish would feel it and back off; if too large, he would slip through. The ideal size would allow the head to get through, but not the body. If he tried to back out, he would be caught by the gills. The best fishing was at night when the fish could not see the net, though in heavy runs they would run into the nets in the daytime. The fisherman went back and forth along his net in his skiff, picking the fish out of the net and throwing them in the skiff. If a fish was tightly caught by the gills, he would soon die of suffocation. In heavy runs, where it was not possible to pick out the fish as fast as they got into the net, enough sometimes died in the net to pull the floats under water, and a net might be lost. Then it was necessary to look for a place where some floats were still in sight, pull the net, fish and all, into the boat.

Maintaining nets was an endless job. Each day it was necessary to go over the net and repair the damage from the previous night. Drift would get caught, submerged snags would tear it. Seals would tear holes in it. These are hair seals, not the fur seals, and are considered unmitigated pests by the fishermen, not only because of the damage they do to the nets, but because of the damage they do to fish caught in the net. They are smart enough to know it is easier to catch a fish in a net than one free in the water. They take just one or two bites from the belly of the fish, spoiling it for sale, then moving on to another.

Defects in the net are repaired by weaving in new cord. The process is like weaving the new net, and some men get very skilled at it. I heard the tale of a young woman whose mother tried and tried to teach the girl to tat, but without success. Her boyfriend persuaded her to help him weave a new net, taught her the technique, and after a time she discovered that was what her mother had tried to teach her, using smaller cord.

At intervals, all nets had to be taken out of the water to get rid of moss and seaweed that got caught in the meshes. This was done by immersing them in a large vat or tank in which copper sulphate, or "bluestone" had been dissolved. This killed the moss, which would then fall off or blow away after the nets had been spread on racks, made of parallel poles, to dry.

Since most of the fish were caught at night, conditions were ideal for theft. Nobody could see all of his net all the time, especially when it was foggy, as it was quite frequently. Nobody knew how many fish his net caught on any given night, nor could he be sure some were taken. If the thief was not too greedy, he could operate a long time. If he got caught, it was more by accident than intent. There were always tales floating around of fish pirates, and probably most had some factual basis. The more reputable fish buyers made an effort to eliminate the pirates, but it was a hard job. If the pirate went through the motions of putting out a net of his own, it was impossible to separate the fish he caught legitimately from those he stole. And some buyers didn't care.

When the salmon run was on, it was necessary for the fishermen to run their nets frequently. Even when it was too dark or too foggy to see what he was doing, one could pull his boat along the cork line of his net, picking out the fish by feel when he couldn't see. This was also the time the fish pirates would pick to rob a net, when the owner couldn't see what was happening.

One fellow, who admitted to me he sometimes made a few extra dollars that way, told me: "When the fish were running good, after two or three days, when the owners of the nets were getting pretty tired, was my time to make a little extra money. I know the bay pretty well, and can find my way in the dark or fog. I'd spot the outer end of a net, or maybe two or three nets, while it was still light, then I'd go there after dark and pick out what fish were caught. I'd keep very quiet and if I heard anyone working along the net, I'd drift away.

"One night I was following the usual procedure, when I felt a motion of the net like a real big fish was in it. I kept working towards this big fellow, hoping I could get to him before I heard the owner coming. Suddenly my boat bumped into something, and when I went forward, I saw I was up against another boat, and the fellow in it was looking forward, just as I was. Then I recognized my next door neighbor. I had suspected he was selling somebody else's fish but was not sure. Now I knew.

"We compared notes, and he thought he felt a big fish in the net, too. Each of us had felt the other, but neither had heard anything, so kept on working until we bumped."

Slim Jensen was a commercial fisherman. One day he told this one.

"One night, while I was drifting on the bay, I got a toothache. It hurt so bad I couldn't keep my mind on my fishing, so, as we drifted past Bay City, I told my partner I just had to do something about it. Old Doc Henson (a semi-retired dentist) lived not far from the Bay City dock, so I went to his house. It was about three o'clock in the morning, and we had left our net in the water; the fish were running good so I was in a hurry to get back. I banged on his door until he came. He had on a light colored flannel nightshirt that was flapping around his skinny old shanks; he was barefooted and had on a nightcap. His long white whiskers were blow-ing in the wind. I told him my trouble and he said, 'Just a minute, I'll fix it.' He went in the house and came out with a pair of forceps in his hand. They were dark in color; I could not see whether from blood or rust, and my tooth hurt so bad I didn't care. He had lighted a kerosene lamp in the front room and the light came out through the open door. As he came out, he was wiping that forceps on the sleeve of his nightshirt. 'Sit down,' he said. So I sat on the edge of the porch. 'Open your mouth. Twist around so I can see.'

"So I twisted around to face the light; he latched on to that aching tooth, and out it came. Oh, how it hurt! I thought it was hurting before, but that was nothing.

" 'Spit!' he said, and I spit, several times.

" 'It will quit bleeding after a while. Charge is one dollar.'

"I paid him and he went back in the house, again wiping that forcep on the sleeve of his nightshirt. I went back to the boat and we hustled out to find our net. Leaving a net like that is just asking a poacher to take over, but evidently none had noticed our absence, for there were several fish in the net."

Sport fishing was entirely different. Salmon were taken on hooks, and various lures were used. Mature salmon seldom feed when they come in from the ocean, so bait is not very useful. All sorts of gadgets are used to attract, most of them a variation of a spinner, drawn through the water behind a slowly moving boat, or by reeling in a line cast out. An enthusiastic fisherman could easily catch more salmon than his family could use, and it was a common sight to see a fisherman going from door to door, trying to give away some salmon steaks. And when fishing was good, people who would accept such a gift were not too plentiful.

Disputes between commercial fishermen and the sports fishermen started when the salmon runs began to dwindle. The sports claimed the commercial fishermen were so greedy they did not let enough fish go up to spawn. The commercial men claimed they were not guilty, that the diminishing runs were due to seals, offshore fishing, contamination of the spawning beds, or obstructions in the streams. I don't know the answer; probably both had some truth, but certain it is that the runs became smaller and smaller until there were not enough fish for the commercial men or the sports.

A few years ago, Oregon outlawed commercial fishing in the bays and rivers. As far as Tillamook was concerned, it was a dying industry, anyhow. Since then, I understand, the runs are gradually increasing and sports fishing is getting much better.

Other fish than salmon are taken by the sports fisherman. Of course, trout are the prime game, but I have no intention of discussing trout fishing; I know from experience I am not competent. But there are a lot of trout streams in Tillamook County and a lot of people catch a lot of trout. There are bass and other fish in the lakes. Every rock, jetty, dock or other structure that juts out into salt water is a spot for optimistic anglers. They catch so-called "trash fish," ling cod, red snapper and others. Some of these are very fine eating; for me, tastier than salmon. Also, I have seen a few people fishing in the surf; do not know how they made out.

In an area where fishing is so general in both fresh and salt water, a certain number of fishing accidents are to be expected, and we got them. The most common was a fishhook caught in some part of the anatomy, most often a finger. Because of the barb on the hook, few people were able to free themselves. Experienced fishermen seldom had this mishap, but children and tourists frequently did. I took out two hooks the same day for one man, a preacher. I have often wondered what he said when the second hook caught him.

Fishermen have a reputation for telling tall tales, and Tillamook fishermen were no different. Tales of being chased up trees or onto big stumps by bulls, by elk, by bears, were not unusual. Any of them could have happened; most of them probably did. Many stories of many happenings I discounted severely. Some, told me by reliable people, are as follows:

"Ed Miller and I were fishing the Kilchis one day. Ed is a big guy, outweighs me by 50, maybe 75 pounds. Also, he is deathly afraid of water; won't even wade up to his knees, but he does love to fish. We came to a place in the river where a little creek runs into the river at an angle, enclosing a triangular point of land. On the creek side, the land slopes gently down to a gravel bar. Just at the mouth of the creek, partly in the creek and partly in the river, is a good fishing hole. From this creek bank, the ground slopes up until it breaks off in a steep bank, maybe ten feet high, straight up and down, to deep water in the river.

“We were fishing this hole, and Ed hooked one, a nice one. He started reeling in, but his line fouled. Somehow, in one of his runs, that fish managed to throw a half hitch around the tip of the pole. The fish was well hooked, and, since he couldn’t reel him in, Ed tried to lead him up onto that gravel bar where he could net him. He kept backing up as the fish got closer to the bar, and before either of us realized it, he backed up so far that he fell over the bank into the river. It was way over his head, but he held onto his pole, and when he came up I was able to pull hard enough on his line to get him up to the bank. As I said, that bank was almost straight up and down, and polished smooth by no one knows how many centuries of river action. He did manage to get hold of a little alder growing in a crack of the rock, just a little tree, not more than half an inch through. It did keep him from being washed downstream, but we both knew it would break if he tried to climb out. I wanted him to let loose and float down stream to where the bank was not so high or steep. I could guide him with the fish line to make sure he didn’t drift too far out. This he flatly refused to do. He was one badly scared cookie.

“I told him to hang onto that little alder, while I looked around a bit. In a clump of willows nearby, I found a pole that had caught there at some high water. By lying flat on my belly, I could get this pole close enough so that he could reach it, but I couldn’t begin to pull him up. He was too big, his heavy clothing was soaked and his hip boots were full of water.

“Pretty soon I saw another couple of fishermen down the river. I called and they came up. Between us we managed to break off a small willow that was long enough. While holding that pole, Ed let us tow him to the lower bank where we could reach him and pull him out.

“Ed had let loose of his rod when we gave him the end of the willow pole. When we hauled in on the line, it was still fouled, so we recovered the rod. Then we looked for the other end of the line and that fish was still on, flopping around in the shallow water over the gravel bar.”

There was one farmer who hated fishermen. He wouldn’t allow one on his place. He had “No Trespassing” signs nailed to every third or fourth post on his line fence. He had a vicious dog he threatened to set on any trespasser. Sam Sutter told me about this situation, then added, "There is a small creek that runs through his place; if it has a name I don't know it. It runs into the Tillamook River, not very far from his property line.

"One day my brother and I were fishing the Tillamook River. It was early in the morning and awful foggy. We came to this little creek and decided to try it. We got one or two before we came to this farmer's fence. There, on a post facing us, was one of those signs.

"We knew that just inside that fence was a good fishing hole. Everybody wanted to fish that hole but was scared of the old farmer. The hole was located in plain view of his house, and that dog could make it down there in about two minutes.

"We stood there, discussing what to do. 'The fog is so thick, he can't see us,' my brother said.

'He might hear us, or he just might be out here looking for his cows. It's about milking time.'

'That hole hasn't been fished and it should hold some good ones. It's pretty close to the fence. I'll bet I can cast into it from here. Would casting into the hole from here be trespassing?'

'I'll bet he'd think so.'

'I don't know, but I'm going to try it.'

"So he made a long cast and dropped his hook into the middle of the pool. Almost as soon as we heard the little splash, we saw the line jerk, and he had a big one on. Then arose the problem, how to get that fish? He could bring it to the bank, but the line was probably not strong enough to lift it out of the water. While we were worrying about that, my brother was playing that fish; and somehow he got into that barbed wire fence. Before we knew what had happened, he was securely tangled up.

"Just about that time we heard a commotion in the direction of the house. The dog started barking, the bull was bellering and the farmer was yelling. We figured he had heard us and was sending the dog. The harder we tried to get my brother loose, the tighter he was caught. It looked like we were in real trouble. About then the dog emerged from the fog, running for all that was in him. But somehow, he didn't seem interested in us. Just then the bull popped out of the fog bank, head down, running like he had important business someplace ahead. Then we realized the dog wasn't after us, he was just trying to get away from that bull.

Pretty soon we heard a crash, and the fence wires went slack. The dog had gone under the fence and the bull had gone through it."

"I jerked brother's shirt open and peeled it off. Then I took my knife and cut away a good part of his trousers, leaving the pieces in the fence with the shirt, and we started to run. After a few steps, brother stopped. 'I'm not going to leave my fish pole for that old so and so,' so we went back. When he picked it up he could feel the fish still fighting the hook. 'I want that fish,' he said, so we carefully stepped over those downed wires, netted the fish, and took off. We had to cut a chunk out of the line for it was tangled in the fence, but we didn't mind that much.

"What we did mind was that mother gave us the devil for the damage to brother's clothes. As I was older, she blamed me. Maybe she was right."

When salmon are in the Bay, a favorite lure for sport fishing is herring. The herring are dead, but they are so fastened to the line that they wiggle around much like a live fish.

One of the local boys had worked in Portland for a time, and while there met and married a typical city girl, one who had hardly been off the pavement. They came to Tillamook just as the salmon were coming into the Bay, and he took her out in a boat with him. The water was unusually clear, so that it was possible to see the fish in the water. He put out his herring, but she did not observe that it was dead. Pretty soon a salmon began to be interested in the herring. He would come close, inspect it, then back away. The herring would again be manipulated towards the salmon. After that happened three or four times, the girl jumped to her feet and screamed, "I think you are a big brute to let that great big fish torment that little bitty one. I won't stand for it, I won't, I won't, I won't." She continued jumping up and down and yelling, and to keep her from putting both of them in the water, he had to quit fishing and head for shore.

A man who went from Tillamook to manage a dairy farm on the rich bottom land along the Columbia River told this tale.

"The Columbia River went along one side of our place. Dykes keep out flood water. A slough from the river cut through my farm. I had seen several fish going in and out the tide-gate and thought that might be an easy way to get a few fish. I made a big cage of chicken wire, with a gate that would let fish in but make it hard for them to get out. I sank this cage to the bottom of the slough, just in front of the tide gate, so anything coming in through the gate would go right into my cage. I had tied a couple of ropes to it so I could stand on the bank and pull it in without getting in the water.

"Next day I went to see what I had caught. I pulled on the two ropes and nothing happened. Then I concentrated on one rope. By pulling with all my might, I finally got one end of the trap out of the muddy water a little bit, and I saw why I couldn't move it. It was literally full of fish; carp, trout, bass, frogs, eels, suckers, all sorts of fish crowded into that cage. Try as I would, I couldn't budge that trap any more. So I got the tractor to give me more power. It proved to be too much power, tore a big hole in the trap. The live fish swam away; the dead ones floated off for the seagulls to pick up.

"I never tried that again. Somehow it didn't seem right."

There were two business men in Tillamook, who shall be nameless, who had adjoining stores. Not only were they competitors in business, but they were always competing as fishermen. With it all they were the best of friends.

One day one of them brought in a beautiful basket of fish. (Catch limits were rather elastic then.) He was exuberant over them, showed them up and down the street, particularly to his next door neighbor. Usually a fisherman is very evasive about where he makes a good catch, but in this case there was no reservation; he told exactly where he caught them.

Next day the neighbor went out. He went to the exact place on the Wilson River, described by his friend, and fished all day with no luck. He surely hated to face his friend and confess his failure. Late in the afternoon, a boy about twelve years old came by. He had a cane pole over his shoulder, and a string of fish so heavy he was having trouble carrying it. Those fish certainly looked good. When he came close, the man said, "Bud, you want to sell those fish?"

"What will you give me?"

He went through his pockets to see what money he had and offered it all.

" 'Tain't enough."

"What more do you want?"

"Gimme some of that fancy stuff you got in that tin box."

He laid out about half the stuff in his tackle box, flies, line, all the stuff a fisherman accumulates.

"You can have 'em."

As he was gathering up his loot, the boy said, "If this keeps up I'll be rich."

"What do you mean, if this keeps up?"

"Aw, there was a dude here yesterday that bought my fish. Seemed like he wanted them pretty bad, at least he paid me a good price for them."

The light began to dawn. "What sort of a looking fellow?"

From the boy's description it was easy to identify his neighbor. When he got to town he went into the neighbor's store, called him into the back room, and told him what he had found out. They had a good laugh at how they had suckered themselves; then one said, "I won't tell if you won't."

They did keep quiet for a time, but the story was too good to keep, and it came out.

Incidentally, the boy grew up, went into business in Tillamook and was there several years. Wherever he is now, if he ever reads this, I'm sure he will recognize himself.

Most of the salmon caught in our area went to the canneries. A little went to the fresh market and some was smoked. Later, smoking was done commercially, but when I first knew the Tillamook country, all the salmon that was smoked was processed in somebody's back yard, each fellow smoking just for his family. There was much rivalry over who could turn out the best product.

Smokehouses were rather simple: small wooden structures, with some provision for a fire in the bottom, and racks for hanging the fish in the smoke. The fish were split into halves, dry salted or soaked in brine, or given some other treatment fancied by the different individuals, then hung in the smoke for the particular length of time preferred. The fish hanging on the racks depended on the skin to carry the weight. Each fellow had his own pet fuel formula, sometimes secret, but I think most of them depended on green vine maple and alder wood.

The product varied from almost raw to nearly board-like hardness. Most of it was medium, dark and dry on the outside.

There are not many things more tasty than a piece of well smoked salmon. One fellow who had never tried to smoke salmon, saw his neighbors' simple smokehouses, and decided there was nothing to it. Also, he checked on the price of smoked salmon, and decided it was a bonanza. He was going to be the smoked salmon king. He built a good sized smokehouse, laid in a stock of wood, and spent all the money he could raise buying a stock of fish, nearly a thousand pounds as I remember it. He cleaned and split them, salted, and did all the rest of what he thought was the proper procedure, hung them in the smokehouse and built the fire. Next morning all the fish was in the fire pit. He had started out with too hot a fire; the fish had cooked, the meat loosened from the skin, and fell into the fire.

It was quite a loss but few sympathized with the guy. He was a sort of know-it-all, so people just chuckled over his mistake.

In addition to fish, clams and crabs used to be plentiful. I recall one time when two of us got two gunny sacks of clams, all we could carry, in less than 30 minutes, from the mud flats off Garibaldi. We could buy a five gallon can of clams for fifty cents. Heavy clamming has so reduced the numbers that the state has now put a bag limit on clams.

There are two ways to catch crabs, nets and traps. In both methods, advantage was taken of the crabs' scavenging habits, by using partially decayed meat for bait. Until prohibited by law, clams were used. Slaughter-house scrap, fish cannery refuse, trash fish, pieces of deceased dogs, cats, or rats, anything sufficiently ripe, will attract a crab.

Crab nets are pieces of small mesh nets, fastened to iron rings, 24 to 30 inches in diameter, so as to make a pouch. A number of these nets, or crab rings as they are commonly called, are used by a man in a boat. A ring is baited, dropped overboard, and the place marked by a float attached to the ring by a rope. He goes forward a short distance and drops another ring. This is repeated until all the rings are out. Then he goes back to the first ring and pulls it up quickly, so that if there is a crab on the net, he will not have time to clamber off. Then the ring is rebaited if necessary and dropped back in the water, and the fisherman goes on down the line. The rings should be pulled about every hour. That is long enough to attract any crab that may be about, but not long enough to let him eat too much of the bait. The number of rings depends on the depth of water fished, and the enthusiasm of the fisherman.

Crab traps are containers, two to four feet in each direction, made of lath, built up like a log house, or of chicken wire over a metal frame. In the top is an opening big enough for a crab to crawl through, but he has trouble getting out. The bait is placed inside and the traps dropped overboard, as with the rings. However, traps have to be pulled only once every 24 to 48 hours, so a man can tend a bigger string.

Traps are used to catch ocean crabs that run bigger than bay crabs. To reduce the labor of pulling the traps from deep water, some boats are equipped with a device called a "gurdy" that is operated by power from the boat's engine.

Deep water crabbing is risky because storms sometimes come up so suddenly it is impossible to pick up the traps. When the storm is over, some or all of the traps have disappeared.

Another hazard of deep water crabbing that interested me is whales. Whales get barnacles, and apparently the barnacles cause itching, because whales like to rub against something. I guess scratching posts are not plentiful in the ocean. They will rub against the lines until they break them.

Hunting, Game and Dogs

Ed Easterly was a state policeman, assigned to fish and game. He had under observation a chap he suspected of spotlighting deer. One night he heard shots in the area where this fellow was accustomed to hunt, and located his car, hidden in some bushes along Highway 101, at the foot of a trail up the mountain side. Rather than undertake a hike up a hard trail in the dark, he parked where he could watch the suspect's car, but where his own car could not be seen from the trail, lit a cigarette and settled down to wait. It proved to be a long wait, and when an empty log truck, without a tail light, went by, he thought it would be a break in the monotony to stop the fellow and tell him the light was out. What he didn't know was that the truck was hot, and the driver was trying to get it to California from someplace in Washington, by driving at night. At the first toot of the siren, the driver panicked, and instead of stopping, gave it the gas and roared off at top speed. That sharpened Ed's interest and he took out after him.

This was during wartime and all cars operating at night on roads visible from the ocean had to have a special dimming device called a black-out light. The illumination from this light was as much below the low beam illumination as low beam is below high. In other words, at best one could see only 10 or 15 feet ahead of him. The chase continued for a couple of miles, both truck and police car at full speed. The truck had its lights on bright—the driver either didn't know of the regulation or didn't care—but Ed was driving by black-out light. After going around a sharp curve, the truck driver stopped suddenly in the middle of the road, turned off his lights and waited until the police car crashed into the truck. The officer was severely injured, but managed to stagger out, arrest the truck driver and handcuff him to his own truck before collapsing.

It was several weeks before I certified Ed as fit for duty. He was ordered to report to the head of the State Police the day before his return to work. We figured he had done something outstanding and was due a commendation. When I saw him the following day, I asked him what the chief had to say. "He gave me hell for wrecking a state car."

Ed was getting along in years, and this injury didn't improve him any. He was finding the fish and game assignment, with its exposure, long hikes and irregular hours, too much of a job, so requested a transfer to traffic, where he would be riding most of the time. The request was granted and a young man was assigned to replace him, a rookie with a lot of steam. Ed was to work with him for a week, showing him the area, the trails over the mountains, the streams most likely to be poached, etc. One day they had taken a long hike over some pretty rough trails, making a big circle back to the little traveled backwoods road where they had left their car. They struck the road about a mile from the car. Ed was tuckered and suggested a rest before walking down the road to the car. After ten minutes, he still was not ready to travel, but the young fellow was restless and evidently wanted to get going. Just then they heard a car coming down out of the hills. To keep the lad busy while he got more rest, Ed suggested that the approaching car be stopped and checked. The young fellow stepped out into the road, stopped the car and started into the routine he had learned in police school.

But Ed was horrified to see that the driver of the car was the wife of a man prominent in state politics. However, there was nothing he could do about it. The young fellow remembered his training and did that car check strictly by the book. He didn't know the driver. To him she was just another woman. He made her get out, took up the cushions and floor mats, checked the numbers against the wanted list, went through the glove compartment, raised the hood, looked under the car, etc. All this time Ed was sweating it out, with visions of both of them losing their jobs. At last the young fellow told the driver to open the trunk and expose the contents. There, under a tarp, was a deer carcass that had been mutilated to conceal its sex, so was presumably a doe, and it was killed out of season.

They took the woman to town, booked her; she arranged bail, etc. After the formalities were completed, the young fellow approached Ed and in an awe-struck manner asked him how he knew to have that car searched. "Boy," Ed replied with a perfectly straight face, "when you have been in this game as long as I have, you learn to play your hunches."

In the mid-twenties, the state legislature passed a new traffic code that was indefinite and ambiguous. As well as I could read it, there was a speed limit of 20 miles an hour for the entire state. That was a bit slow, even for the cars and roads we had then. One day I asked Happy Dunn, the state policeman who had charge of our district, just what it meant. His answer, "Doc, I don't understand it any better than you do. As far as you are concerned, stay on your own side of the road, use reasonable care, and if you think you have a reason for hurrying, I can't see your speedometer. Who knows, maybe one day you will be coming after me." He thought for a minute, then said, "When you do, drive like hell."

John Swanson was an enthusiastic (and successful) deer hunter. The closed season was a dreadful restriction to him, and one he did not always observe. He operated a dairy farm in the north end of the county. When he had occasion to leave the farm, he traveled in a battered old pickup, and his rifle always traveled with him, resting on an inconspicuous bracket in the cab.

One day he had business in the other end of the county. While driving along a lonely wooded stretch of highway, a nice buck started across the road in front of him. The temptation was too great; a quick glance ahead and in the mirror revealed no other car in sight. He grabbed the rifle, hopped out, and before the buck reached the other side of the road, he had a bullet through his head. Before the buck stopped kicking, he was dragged off the road into the bushes and his throat slit. In a few minutes he was dressed out and hidden under some brush. A few handfuls of dirt hid the blood spots on the road.

On the return journey, John stopped opposite where he had hidden the carcass, and pretended to be fixing a tire, until there came a time when no car was in sight. Then he quickly threw the carcass in the box of the pickup, covered it with an old tarp, and headed for home.

In one place, a spur of the Coast Range leaves only a narrow gap along the ocean, and as he rounded the curve approaching it, he saw the state police had set up a road block. There was no going around them, for there was no other road. Besides, he figured the police had seen him and if he turned back, they would be after him to find out why. There was nothing to do but face it, so he drove up to the officers and stopped. They made him get out while they gave that truck a thorough going over. From their conversation, he inferred they were looking for a stolen truck, and his vehicle fitted the description. They looked behind the instrument panel, checked the numbers, took out the floor mat and seat cushions, looked over his rifle, opened the hood for a look-see at the motor, even crawled under the truck. But nobody thought to look under that old tarp. After about fifteen minutes, they decided they had not found the truck they wanted and told him to drive on. That he was glad to do, for it would soon be milking time.

He brought me some of the steaks and they were good. It was a nice young buck.

The same fellow one day saw a deer off in the field, by a clump of blackberry bushes. He stopped, got out of his truck, and shot. Just as he shot, another deer walked out from behind the berry bushes. He killed both deer with the one shot. That was a lot of meat for a small family, but he put both carcasses in his trunk and went home. His wife was very much put out; she wouldn't believe the second deer was an accident. She knew him pretty well. She really gave him the devil for killing so much meat at one time, but finally cooled off and helped fix it for the freezer. Took nearly all night.

Later, this same man decided to quit dairying, and got a job in a sawmill. The work was all right, but when the deer season was getting close, he began to get "antsy." The mill had just gotten a big rush order from a new customer whom they wanted to please, so had announced no vacation for deer hunting that year. To give it force, they also announced that anybody taking time off without leave during hunting season need not come back. That was tough on a confirmed hunter; he wanted to go hunting, but he didn't want to lose his job. He worried over it so much that his wife took pity on him and arranged for a friend in a distant part of the state to send a telegram: "Jim not expected to live, wants to see you."

This telegram was delivered to the house during working hours, so the wife took it to the mill office. (She wanted to go hunting, too.) As he told me the story, "The boss called me in and said, 'Of course, under these circumstances you can have all the time off you need.'

"I didn't know anything about this telegram, so had no idea what he was talking about, but had sense enough to keep my mouth shut; just said "Thank you' and walked out. I was working the night shift, and by the time I got home, it was good and dark. We put the camping gear in the pickup and covered it carefully with a tarp so nobody could see what we were taking, and didn't buy any groceries until we got out in the valley. The weather was perfect, and I shot the biggest buck I ever saw in my life, with the biggest set of horns I ever saw. And I knew I didn't dare brag about it, for I didn't want the story to get back to the boss. I didn't dare bring home the hide or the horns, and I cut the meat into small chunks. We timed it so we would get home about three in the morning so nobody would be likely to see us carry the meat or camping gear into the house.

"When I reported for work the next evening, the boss asked how Jim was, and I told him 'Much better.' It was some job keeping a straight face and I had to tell a lot of lies to the boys around the mill, too. But the worst was not being able to brag about that big buck. And many is the time I have wished I had those horns.

"The real payoff came later. When we cooked some of the meat, it was so tough we couldn't even chew the gravy. We ground up some of it into hamburger, but couldn't eat it. Even the dog wouldn't eat it. Only the chickens could. Our chickens had the fanciest food any chickens ever had while that meat lasted. I'd like to know how old that buck was."

Ed Samuelson was a big strong man, big frame and plenty of upholstering. He walked with a limp from a World War I injury and always carried a heavy cane. His farm had a lot of hill land, only fit for grazing, and he ran a bunch of beef cattle.

One day it dawned on him that he had not seen a number of his heifers for several days, and decided to walk the fence to see if there were any breaks in it. It was a warm day; he had gone mostly uphill, and he was not a good walker anyhow. Soon he was tired and stopped to rest in the shade of a tree on the edge of a patch of brush. He stood there quietly, trying to decide whether to go on or turn back.

While he was debating with himself, a spike came slowly around that patch of brush, browsing as he moved along. He was almost close enough to touch. I don't know which was more sur-

Elk sometimes made themselves right at home.

prised, but Ed moved first. He hit that deer with his heavy cane so hard he broke its neck. Also broke the cane! That is the only time I ever heard of killing a deer with a club.

Ed had a little difficulty with the game warden over that deer, but was finally able to convince him it was self-defense.

Dale Perkins had a ranch up the Trask River. The Trask River Valley is narrow, with high wooded hills on either side. Every ranch there borders the river on one side and reaches to the hills on the other side. Early in the summer, Dale saw a three-point buck and two does with fawns in his pasture. He saw them there nearly every morning when he went out to bring in the cows. At first they were afraid of him and would go over the fence and into the timber as soon as he came in sight. As they became used to him, and he never paid any attention to them, they became so tame they would let him come close to them and would not run until he began to yell at the cows to get them started.

As it got along to September, Dale thought he knew where he was going to get his buck. Only bucks could be shot then. On the first day of the hunting season, he got up long before daylight, made a big circle up the hill to get back of his place, came down the hill to the edge of the timber to wait for daylight. When it was light enough to see, there were his cows, most of them lying down and chewing their cuds, and there were the does and fawns, but no buck. And he did not see him again until the hunting season was over! Then again he came down to graze with the cows. This time Dale threw rocks at him.

How did that buck know when hunting season started and ended?

Ranging on the ridge north of the North Fork of the Nehalem River and north into Clatsop County was a herd of elk. There is a lot of feed in that area and this herd increased rapidly until it became quite a pest. The elk are so big they can walk right through a barbed wire fence, with serious consequences to the fence. They seemed to get a lot of fun out of chasing cows, and the cows are afraid of them. The laws forbid shooting them except for a few days during the open season; then only if you have an elk license. So it is easy to understand why the ranchers in that area didn't like elk.

One morning one man came out on his back porch, on his way to the barn, when he saw elk in his garden, one eating his strawberry vines, another his beans; another was in the barnlot eating from a pile of hay; three or four more were in his pasture, and no cows in sight. That was too much. He stepped back into his kitchen, picked up his rifle and shot three times, getting the two in the garden, and the one in the barnlot. Leaving them where they fell, he called the game warden. The warden came, got the carcasses and filed charges against him for killing game out of season, but the case was thrown out of court. The judge knew something about ranching.

One fellow, a disabled veteran of World War I, gas-damaged, was a confirmed hunter. He couldn't walk very fast, but he was persistent and probably knew the hills of north Tillamook County as well or better than anyone else. When World War II came along, he was caught with only a few shells for his rifle, and he couldn't get more. After studying over the problem, he decided the way to get the most meat for his family from his three shells, was to kill an elk. So he very carefully hoarded those three shells. Opening day came, and he was in the woods early. There was no trouble locating a herd. He had watched them so much the previous few weeks that he knew about where they would be. A careful stalk got him close; couldn't take a chance on a long shot with only three shells. He picked out the animal he wanted, balancing quality of meat against quantity. When he was close enough he couldn't possibly miss, he stood up, aimed, and pulled the trigger. Nothing happened. He pumped another shell into the barrel. Nothing happened. And the third one. All three of those shells were duds. The elk stood there and looked at him, just like a herd of cows. He was so angry he clubbed his rifle and started to run toward them. Then they trotted leisurely away.

Just off the coast near Oceanside is a group of rocky islets that since 1907, have been a federal wildlife refuge. They are inhabited by sea lions and birds, mostly murres. These birds lay their eggs on the bare rocky ledges. The reason they do not roll off is that they have a sharp pointed end, so if they roll, they roll in a small circle. The birds look somewhat like little penguins. Between these rocks and the mainland is a spectacular display of sea anemones and other underwater growth. One of the long-time residents of Tillamook told me this one.

"When I was a young fellow, we used to go over to the coast above what is now Oceanside, for target practice. We would sit on the edge of the cliff and shoot at things over on the rocks. That was before the government took it over. There was always something over there to shoot at. One Sunday afternoon, Joe Ely and I went over to sharpen up our shooting for the deer season. When we got set, we noticed a moving white spot that we had never seen before, so we both shot at it several times, but we missed. After we had each shot maybe half a dozen times, the spot jumped in the water and swam ashore. Then we saw it was a man with no shirt. When he reached the shore, he took off down the beach as fast as he could run. We were pretty scared, thinking about how we might have killed him. Too bad we missed, though. It would have saved the county a lot of money. He was a wanted man, who had been hiding out over there, and he raised a lot of disturbance around the county before the officers finally killed him in a gun fight. We could have saved the sheriff's men a lot of trouble if we had been better shots."

The opening of deer hunting season was a sort of annual madness. A large part of the men, and many women, took part. It was talked about and prepared for weeks in advance. All sorts of supplies were laid in, food, ammunition, camping equipment, etc. The quantity of whiskey bought was surprising. I never could understand how they could drink that and hunt, but maybe it is easier to hit one of them if you see three deer at one time. So many people went hunting that the mills and camps closed down for a week. They knew they would not have enough crew to operate; besides the bosses wanted to go hunting, too.

Some people hunted near home, some went outside. Everybody knew just the perfect place to hunt, and everybody had a different place.

The areas around Tillamook are so brushy and rough, that a deer is in sight for just a moment, and some people got so quick on the trigger that they sometimes shot when they should have waited. I recall one man and his teenage son that were hunting together. The son made a large circle to drive possible game towards his father. As he was approaching where his father was hidden, he had to scramble over a large log. The father thought he was a bear and shot him.

Another fellow had killed a buck. To carry it out, he got a foreleg over each shoulder. His brother saw the antlers over the brush, shot where he thought the neck would be, and got his brother through the head.

Such incidents happened nearly every year. They were so common, I decided the best place to be during the hunting season was not in the woods.

Then, as now, the once a year sportsman was a problem. He often got the real sportsmen into trouble, as well as getting into it himself. It is the ignorant and inconsiderate actions of a few such people that have caused the feud between sportsmen and landowners, and has closed so much territory to hunting and fishing.

Duck hunting in the early days was wonderful, if you liked duck hunting. As far as I was concerned, wading out on the tidelands before daylight, then huddling in a cold, wet blind until some ducks took a notion to fly over, was just one jump from insanity. You could buy the ducks for one dollar a dozen, if you wanted ducks. But if you liked duck hunting, it was fabulous. There were lots of ducks, and there was no bag limit. What more could a crazy duck hunter want?

One such hunter told me, "I always load my own shells. One day I was loading some when my wife called me to dinner. After dinner I went back to my job, picking up where I left off, but I guess I forgot where I left off putting in powder. Anyhow, one shell got an extra load.

"I had a dandy setup. The last winter's storm had washed a big pile of brush against a large drift log, right under where the ducks often fly. I had cut out a hole between the brush and the log, big enough to get into, with enough brush over me to hide me, but not enough to cut down visibility very much. To top it off, there was a branch on that log that was just the right height to sit on. I could sit on that branch, out of the wind, and wait for the ducks to come. When they did fly over, I'd stick Old Scatter-gun out through the brush and shoot without ever getting up. I had a good dog and he'd bring 'em in. Could you figure anything sweeter?

"The morning after I was telling you I got mixed up on my shell loading, I was in my blind before daylight. When the ducks started coming over, I began banging away, and pretty soon my dog had several ducks piled up just outside the blind. Then I shot once more, at a duck directly overhead, and I used that double loaded shell. There was a tremendous bang! I was forced down hard on that limb I had been setting on, and it broke. That let me down full length in the mud. Somehow or other, in falling, my legs went under the brush pile and caught there. All my efforts to pull them out just made my body slide back and forth in that slippery, stinking mud, until I was in so deep I had a hard time keeping my face out. I knew I had to get out of there before the tide came in, and I guess I panicked and thrashed around like a fish in a net. All that did was to push me in deeper. At last I took time to look things over carefully. By bracing my right hand against the log, and grasping branches in the brush pile with my left, I managed to pull myself to a sitting position, though my thighs were several inches under the mud.

"By feeling along my legs and breaking away the smaller branches I determined that both ankles were lying on what seemed to be a small log buried in the mud. On top of my ankles, was a branch about four inches thick that was on the bottom of the brush pile. This branch was completely buried in the mud. Evidently in falling my toes had gone between the log and branch; the force of the fall had lifted the branch enough to let my heels through, and the branch had snapped back, trapping me very neatly.

"I tried to slide my feet out of my boots, but couldn't. I tried to use my gun barrel for a pry, but could find nothing solid for it to rest on. Then I thought of my knife. It was in the scabbard on my belt, and fortunately hadn't fallen out in all my floundering around. I started hacking away at that branch, working in the mud, entirely by feeling. That branch was some awful hard wood, and waterlogged, but I kept at it. By the time the tide had just reached me, I got that branch cut, and had one foot released. I dug that into the mud until I had it firmly against the little log my ankles had been resting on, and that gave me something solid to push against, so that I was soon able to get the other foot loose.

"By this time, the water was four or five inches deep, and my gun was entirely covered, but I felt around until I found it. It was completely filled and coated with mud. I was plastered solid. I waded out and went to a little creek nearby, laid down in it and scrubbed the worst of the mud from me and from the gun. My wife wouldn't let me in the house until I peeled off every stitch of clothing. It was sure a job to clean up that gun.

"I still hunt ducks, and sometimes use that blind, but it isn't as much fun as it used to be."

Another man that was an avid duck hunter was Sam Slocum. Every possible chance, he was out in the wet and cold, looking for ducks. He always carried an ample supply of snake-bite remedy, for he was terribly afraid of snakes. Of course, he knew no poisonous reptiles lived west of the Cascades, but he was afraid the snake he might meet would be a tourist and not familiar with local rules. So he always carried the medicine. He had a theory that when the snake bit, it might be too late to take the medicine, so he took it frequently and liberally, to be well prepared if the emergency arose. Strangely enough, the snake-bite medicine never seemed to affect his shooting eye, for he consistently brought in game, but sometimes it did seriously affect his judgment and his locomotion.

One day he came home after a long session on the tide flats, more than usually affected by the medicine. He had a shell in the barrel of his gun, and when he tried to get it out, it jammed, and he couldn't move it. After trying for quite a while, with no luck, he thought that if he fired that shell, the ejection mechanism of the gun might kick out the empty shell. So he went into the bathroom, put the muzzle of the shotgun in the toilet and pulled the trigger. The gun kicked out the empty shell all right, but the bullets knocked the bottom out of the toilet bowl and drilled a hole in the floor.

For months, when the fellows that knew the story met him, they would pretend to aim a gun downward and yell "Boom." As far as I know, none of this abated his interest in duck hunting, or his addiction to snake-bite medicine.

Another story that strains credulity a bit concerns the enormous number of ducks that formerly passed through Tillamook on their semi-annual migration. One of the oldtimers declares that he once fastened a wire to a rifle bullet, fired it through a flock and had ten ducks strung on the wire!

Another chap told me he was standing on a hill one day when a large bunch of snipe flew up the valley, just about on a level with him. He fired both barrels of his shotgun at once, but his aim was a little low, no snipe fell, but he did pick up a bushel basketfull of snipe legs.

When I first went to Tillamook, ducks were unbelievably plentiful. One man told me he once killed 17 mallards with three shots, standing in his barn door.

Another regularly lost several acres of pasture land every year. For some reason the ducks liked to spend the night in one corner of his field. They tramped out much of the grass and there were so many feathers and droppings that the cows would not eat what little was left. He tried shooting to scare them away but he could not keep from hitting some; then he would have to gather them up and bury them to prevent a stench, for the prevailing wind was towards his house. The smell of the droppings was bad enough. He tried all sorts of other schemes to make enough noise to scare them away, but nothing worked more than one or two nights. At last he just gave up. Advancing civilization solved the problem for him in the end. Last time I talked to him, he said sometimes a few ducks would spend a night in the old resort, but not many.

Hunting dogs were not too common in the Tillamook country, but there were some. There were two kinds, bird dogs and cougar dogs. Not many people hunted cougar, but there were three or four who did. Each of them kept one or more dogs, for hunting cougar involved finding a trail and following it. For that purpose,

dogs were essential. I think cougar dogs were generally some type of hound. The owner trained the pups by running them with an experienced dog.

Bird dogs were used on water fowl. Many hunters got along without them, but when available, they were a big help. Most hunters picked up a likely looking pup and trained it themselves. I saw collies, shepherds, scotties, and just plain dogs that did adequate jobs of retrieving. Can't remember seeing a full blooded retriever.

One fellow, who had a small shepherd, claimed he was a good retriever, except for one fault. "About three years ago, Spot and me were working along a slough of the Tillamook River and we came across a Canada goose. He evidently heard us before we saw him, for he was leaving, fast, almost out of range. I took a quick shot and broke a wing. He fell in the middle of the slough, and before I could stop him, Spot went after it. He was young then, and had never encountered a wounded goose. He had handled a lot of wounded ducks, and evidently thought this was just a big duck. He soon found out he was wrong. When he got close enough, that goose grabbed him by the nose and held on.

Spot couldn't even howl; every time he would try to get his head far enough out of the water to howl, that goose would jerk him down. The goose was almost as big as the dog. Of course that broken wing handicapped him, but he was a better swimmer than Spot was, and he could stay under water longer. After a while Spot got loose and headed for shore, that goose chasing him. He made it to shore, blood spurting from his cut nose, and headed for home as fast as he could run. When Spot was out of range, I fired again and killed the goose. The wind soon brought it to where I could reach it. It sure was a fine goose.

"After that experience, Spot would retrieve anything except a goose. I shot a few more while he was alive. He would tuck his tail between his legs, look at me and whimper, but I could never make him go in the water after a goose, even if it was dead. I have put a dead goose on the ground before him, poked it, pulled feathers out, turned it over, to show him it was harmless. He would watch from a few feet away, but he wouldn't come close. He wouldn't even eat the goose bones we threw out."

A wealthy man out in the Valley bought a large farm in Tillamook, stocked it with blooded cattle, good machinery, built a fine big barn, house, etc., and put his son in charge. The young fellow was about 30, not a bad fellow, but just a rich man's kid. He had grown up with a bunch of similar kids with too much money, too little discipline. Don't think he had ever done anything useful in his life. His idea of running the farm was to walk around and tell the hired men what to do, even if they knew more about it than he did.

He tried to be a good fellow with his neighbors, but just didn't know how to get along with men that had to work for a living. He posed as a sportsman; he bought guns, fancy clothes, fishing equipment, gadgets of all kinds. He had everything anybody suggested but he got less fish and game than any of those he went out with. He noticed the miscellaneous collection of mongrels his neighbors had trained to retrieve, and decided he must have a hunting dog, and it must be a real hunting dog. So he sent to a famous kennel, bought a high priced dog and paid $175 to have him trained. When he took that dog hunting, all he would do was chase rabbits. To go in the water, or to look for ducks or quail in the weeds—nothing doing.

One of the problems of all livestock men is that of dogs running or killing their stock. This was not a big problem in Tillamook County, but it did sometimes come up. Sometimes it was dogs that had gone wild; more often it was dogs that were somebody's pets in the daytime but different animals at night. The sheep men had many animals killed, but I never heard of a cow getting anything worse than a few nips. Apparently the dogs just chased the cows for the fun of making them run. However, the dairymen did not look on it as just good clean fun for the dogs. A cow that was chased at night would not give much milk the next morning. That was an important matter.

One summer there was an unusual amount of complaint about cows showing evidence of being chased. All around the central part of the county there was complaint. Finally an organization was formed to meet the problem. The area was divided into districts and a schedule of patrols was arranged. Every night a team of two men would travel the district. As far as it was possible, cows were pastured at night where they could be seen from the road. In this way every herd could be seen several times every night.

The head of this organization was a well known man who had made many complaints. His cows had been chased several times, and he was particularly bitter in his denunciation of people that allowed their dogs to run at large. He even circulated a petition to the legislature to pass a law declaring anyone who harbored a dog that chased stock was guilty of a misdemeanor and subject to all the penalties pertaining to a misdemeanor.

One night when he and his partner were on patrol, they heard a disturbance in a field near the road. They got out of their car and quietly worked their way to where they could see what was going on. Sure enough, it was a group of dogs chasing the cows. They waited until they had a chance to shoot, and dropped the leader of the pack.

When they investigated what they had killed, they found it was the leader's own dog, one he thought was a lazy, inoffensive, worthless animal that was the pet of his kids. Now he knew why the dog laid around and slept all day; he got his exercise at night.

He decided not to present his petition.

Prohibition

During prohibition time, Tillamook County was no more law abiding than most communities, probably not as much as many, because of its high percentage of people of foreign birth, who were used to using alcohol regularly, and who did not take kindly to being deprived of it. As a result, there was a ready market for liquor, and where there is a market, some enterprising American is going to supply it. In fact, bootlegging became almost respectable. The irregular coast line made an ideal place to land imported stuff, and a considerable quantity probably was landed. My impression is that most imported liquor was intended for big city markets, and little went to local trade.

One of the local men was walking the beach near the south end of the county, the morning after a severe storm, when he found 150 cases of imported Scotch. Like a good citizen, he called the law and turned it over to them. That was a real test of good citizenship, for it could have been sold for ninety dollars a case. There was much debate as to whether he was really that good or just dumb.

Local bootleggers were numerous; some were prosperous; some got caught. Some that got caught were still prosperous. Several were tempted to try their hand at manufacturing. Many of these, of course, came to grief, for the local officers did their best. There were a lot of problems in the business of illegal manufacture of liquor; where to get the grain without arousing suspicion, the telltale odor of fermentation, unusual activities attracting notice, getting rid of the spent mash, the desire for too much profit, etc. The chances were good that a moonshiner would ultimately get a tit caught in the wringer.

After it was all over, I learned of a still that had operated continuously for a few years within a mile of the courthouse and within 50 feet of a well-traveled road. A young Swiss couple, Chris and Bertha, operated a dairy farm. They were a pleasant young couple, son and daughter of local dairy farmers, who as youngsters took an active part in school and community affairs. When they married, they rented a small dairy farm; she ran the farm and he worked in a logging camp. When they accumulated a little stake, they rented a larger farm, bought more cows and increased their income. They repeated this maneuver two or three times, ending up on a very good farm near town, with enough cows that the man could quit his job in the woods and devote his entire time to the farm. They were well known and respected as a hard-working, up and coming young couple, who minded their own business, lived quietly, and participated in community affairs. They operated their ranch by their own labor, never hiring any help.

An uncle of this young man had worked in a distillery in Switzerland, and with his help and advice, Chris assembled the equipment for a still and put it together in his barn. The heating plant was bought as an electric hot water system to enable him better to clean his dairy utensils. Water was available from the usual barn system, for watering the cows and cleaning up. The mash vats were made from lumber bought a little bit at a time at various lumber yards, and they were vented into his silo (as was the still), so the odor never caused any comment—it was too much like the normal silage odor to be noticed. Grain was bought from half a dozen feed stores within 50 to 75 miles, no two in the same town; never buying at one time more than was reasonable for the number of cows he was known to be milking. He picked up this grain with his own truck, so suppliers never met and compared notes. This aroused no suspicion, for a number of dairymen shopped around for feed bargains. The spent mash he fed to his cows—it was good cow feed.

He put his product in fruit jars, which could be picked up, a few dozen at a time, at any grocery store, buying enough during the canning season to last all winter. He paid cash for this outside stuff, though he had a substantial bank account and paid all local bills by check. Providing yeast was no problem to a woman that had made bread since she was a small girl, especially if she had the proper automatically regulated electric gadgets for maintaining uniform growing temperature. Electric heat made it possible to operate the still continuously without constant supervision and without smoke. The unusually large amount of electricity used was explained by openly buying all sorts of electric appliances, including a heating system for the house, so the power company never questioned the amount of juice used. The product was sold for a reasonable price, only to people known to the couple. And that was no problem, for they had lived in the community all their lives, belonged to many of the community organizations, and knew nearly everybody, or at least all the oldtimers. They knew who could be trusted to keep his mouth shut, and who couldn't. They never made a sale at their home, so no suspicious characters would be seen coming and going. They never tried to get too big, being satisfied with what they could handle by themselves.

Chris had to quit because gasoline rationing of war time cut down the use of his truck. He could not run around the country to get grain, and he could not buy enough locally without attracting attention. Also, the grocers were beginning to ask questions about so many fruit jars.

All of this I never suspected, in spite of the fact that I did much work for them and thought I knew them well. How they were able to do it, I couldn't see, for the farm was big enough to take the entire time of one couple, and they ran it well. Chris told me after he had quit that they made more off the still than they did off the cows "and the cows paid good, too." The war was still on when he told me this, and he was not sorry to have a rest. Just running the ranch seemed like a vacation. He said he had dismantled his plant and had the essential parts hidden. Asked if he expected to start it up again when rationing was ended, he said, "Maybe, but the kids are getting old enough to begin asking questions. On a farm it's pretty hard to put anything over on the kids when they get big enough to be nosey. And, I'm not very proud of it. It paid good, but I might get caught, and I don't want my kids mixed up in that kind of a mess."

One of our more active bootleggers was Alec Swenson, who, as a cover, operated a small dairy farm in a little valley several miles up in the hills. He was so brazen about his bootlegging that he ran around the country in a robin's egg blue Chrysler, the biggest, most expensive made, so flashy it could be recognized as far as it could be seen.

He ran up a bill with me, $20 or $30 as I remember it, and he ignored all statements. At last I got tired of wasting postage on him and turned the account to a collection agency. A few days later, I met the deputy sheriff for our area, a good friend of mine. When he saw me he started to laugh.

"You know that bill you had against Alec Swenson, the one you gave to the collection agency! Well, they got an attachment on that big blue Chrysler and sent the papers over here for me to serve. By the time they added court costs, mileage, etc., the bill was more than doubled. I took a driver with me and went up to his place. There was the blue car in the barn-lot. I walked over to it, and the key was in the ignition lock. I got in and started the motor, and here came Alec out of the house, roaring like a bull. When he got quiet enough to listen, I handed him the attachment writ and told him I was taking the car to the county seat.

" 'But you can't do that. I need that car in my business. It is a tool of my trade, and you can't attach the tools of a man's trade.'

"My reply was I didn't know whether I could do it or not, but I was doing it, and that paper from the court was my authority. Either he paid the bill or I took the car. He yelled and cussed and threatened, but I just started to drive off. He ran after me, and I let him catch up. He said, 'If that is the way you feel about it, I guess I'd just as well pay it.' Then he pulled out of his pocket the biggest roll of bills I ever saw outside of a bank. He peeled off three or four bills and said, 'That will more than cover it. Keep the change. I can't afford to let you take my car.' "

You would think an episode like that would make him mad at me, but it didn't. He, and his children, after they had families of their own, stayed with me as long as I was in Tillamook County, and always paid promptly.

Later, Alec told me this story.

"During prohibition time I had a still out in the hills back of my house, half or three quarters of a mile from home. I used a kerosene stove for heat; no smoke, and I wouldn't have to visit so often. One evening, just before dark, I started to the still carrying a five-gallon can of coal oil. I wanted to get back before dark, for it wasn't a very good trail, so was hurrying and not keeping a close lookout. The trail was across my pasture where there were several clumps of blackberry bushes. As I hustled around one clump, I ran smack dab on a mother bear and two cubs, eating berries. One of the cubs was on the path and I nearly stepped on him. He was so close that when I dropped the can of oil, I guess

I dropped it on his paw, for he bawled like a baby. Mama bear evidently thought I had designs on her offspring and came at me with a roar. I didn't wait to argue, just ran, with Mrs. Bear one jump behind. The nearest tree was a little alder, but it looked better than nothing, and that bear was gaining on me, so up I went, as high as I could get. But that wasn't very high. Mama bear was on her hind legs, clawing and slashing at my boots, and missing them by inches only. This went on for some time, and she showed no signs of getting tired of the game, but as far as I was concerned, it was a bit monotonous. I had no weapon, had even lost my pocket knife the day before. At last I thought of my snoose can. It was nearly full. So, when she gave a big roar, I reached down and emptied it in her mouth. She looked surprised, dropped to all fours, started coughing and clawing at her face, then headed for the creek, making noises that sounded like Ulp, Ulp, Ulp! She didn't even wait to call the cubs, but they followed. I climbed down, retrieved my can of oil, went on up and tended to my still. Never saw that bear again and didn't want to—afraid she might remember."

Bootlegging activities in the area received a severe setback sometime in the early twenties—I cannot give the exact date.

One day a couple of strangers came in on the train. They put up at the hotel and for several days circulated about town, talking to the older people, asking a lot of questions about Tillamook County, its history, industries, resources, customs, peculiarities, etc., and taking copious notes. They also hired the local taxi for days at a time, and rode all over the county. When questioned about what all this was for, they were very evasive, gave some sort of vague talk about a book on the Oregon coast area. After several days, they had talked to so many people that nearly everybody knew about them and was curious. Then they approached the president of the Chamber of Commerce and asked for the privilege of addressing a membership meeting of the Chamber. They said they had an announcement that would be of general interest. A meeting was called, and nearly every member was there, including representatives of all the newspapers, not only those in Tillamook, but also those in Cloverdale, Garibaldi, Wheeler and Nehalem.

The men were properly introduced by the president and given

the floor. The spokesman said all the usual things about a pleasant visit to the community, friendly reception, beautiful country, great possibilities, etc. He spread it on thick. Then he came to the reason for the meeting. He and his associate were scouts for one of the big movie producers, which one he was not at liberty to divulge, but he assured them it was one of the big ones. The company had been offered a script for a story located in Tillamook County, a script of tremendous possibility, maybe one of the big dramatic hits of the decade. To achieve the maximum impact, a great deal of authentic local color was necessary, logging camps, dairy farms and commercial fishing. Their task had been to see if it was available, also if accessory facilities were available. They had found scenery and background possibilities very satisfactory. Enough carpenters, mechanics, painters, etc., were around; machine and carpenter shops were adequate, electric power in sufficient amount was available. Vehicles for hire were in short supply, but they could be obtained in Portland. All in all, their findings were very satisfactory except for one thing, housing for the staff. They would need to bring in about 250 people, actors, technicians, clerical staff, etc., and many of them would want to bring their families. Shooting would probably take at least six weeks. There was considerable housing for rent in the beach areas, but they were pretty well taken in the summer; besides the company did not favor having their people so widely scattered. Could this group put on a campaign to get enough people to take paying guests into their homes to handle this work force, also help sign up five or six hundred people for extras?

After some discussion, it was decided it was a suitable job for the Chamber of Commerce, and they would undertake it. The speaker thanked the group and said he and his colleague would make their report immediately. They left on the train next morning.

In about a week they were back. They said the directors of the company were enthusiastic over the report and had sent them back to push preparations. They leased a tract of land, where the County Hospital now stands, on which was an unoccupied house. The house was fitted up as an office, and a crew of carpenters put at building a set. In front was an Indian village, and behind that a logging camp. Back of that was a bunkhouse. Some of the carpenters questioned why it was necessary to make that bunkhouse so strong, but they were told some pretty rough and tumble fights were to be staged there, and it would be ridiculous if an actor fell against a wall and knocked it down.

At last the set was built, the extras signed up, the housing listed. These activities made every person in the area aware that something big was coming off. Then it was announced that the company people would arrive on a certain day, on a special train, direct from Hollywood. The star was one of the best known glamour girls of the silver screen, the leading man one that caused heart throbs all over the world, the director a man who had handled some of the biggest spectaculars of history, and there would even be a well known horse. However, to keep the competing companies from getting vital information, it was necessary to keep their names secret until they stepped off the train in Tillamook. The special train was scheduled to arrive in the late afternoon or early evening, and a big welcome at the station was to be followed by a party on the set that would get headlines in every newspaper in the United States and maybe abroad—publicity for the movie.

Word was quietly slipped to every known bootlegger in the area that the big party would need barrels of hootch, and if they wanted to get in on the business, to contact Gussie Jones at the Tillamook Hotel four or five days before the party. Most of them bit. The build-up had been so open, and everybody was so familiar with the project, that few of them even considered the possibility of a trap.

About a week before the party, the said Gussie Jones came in on the train, registered at the hotel, and made himself very conspicuously inconspicuous. Most of the time he was in the hotel lobby, but sometimes he walked around, seeming to favor back streets and vacant lots.

When a local bootlegger would appear, he would greet him briefly, tell him to come to Room 241 in fifteen minutes, and knock three times. Also, he must be sure no one saw him come in, and he was not to use the elevator, for the boy might remember. All this was very impressive.

When the bootleggers knocked at 241, Jones would peek out cautiously, identify the caller, look up and down the hall, open the door just wide enough to let the man through, then quickly close and lock it, and close the transom. Then he came close to his guest and said in a stage whisper, "Don't talk too loud." Then would follow a confidential talk about how he, Gussie Jones, was here to arrange for the drinks for the big party, and it would take a lot, for some of those movie people could drink like a camel, and how he, Gussie Jones, could get all he needed in Los Angeles but he wasn't getting it here and he was getting worried, and how much could the guest scrounge up before the party time, and was it good stuff? The price he was authorized to offer, cash on delivery, made a back country bootlegger's eyes pop. When quantity and other details were arranged, he was told to bring his merchandise to the shaded side door of the house on the set, where some trees made it pretty dark. He was told to be there at precisely a certain time when the sheriff and his men would be sucked out of town on a false alarm.

Each bootlegger was given the same treatment, except that each had a different delivery time. All this was scheduled for the night before the party. On the night of delivery, there was no light in the house, but when the hustler drove up to the appointed place, a man, with his hat pulled low over his eyes, but easily recognizable as Gussie Jones, would say, in a low voice, "Just open the door and set your jugs on the floor on the far side of the room. I'll keep watch. As soon as you unload, I'll pay you."

When the vendor came through the door into the dark room, with both hands full of jugs, he would feel a gun against his ribs and a voice would say, "just hang onto them jugs, buddy, and keep walking." A flashlight would show the way out the back door, and along the path to that sturdy bunkhouse.

While this was going on, his car was being moved out of sight to clear the scene for the next appointment. This was repeated every thirty minutes for several hours. The bunkhouse was full of bootleggers and the accumulation of illegal liquor was one of the largest on record. The T man that organized the scheme said it was the most successful trap ever pulled off on the Coast.

And Tillamook was almost dry for a few months, probably longer than any time before or since.

Human Oddities

I presume every isolated community has its quota of odd individuals, but it seemed we had more than our share. One day I was called to see an elderly couple. I looked over the man first, then started to write him a prescription, but he stopped me, saying, "No use to do that, doctor, I never take medicine." The wife broke in with, "You just go ahead and write for that medicine, doctor, and *I'll* see that he takes it."

Then she was checked over, and when I started to write a prescription for her, she stopped me with, "No use to do that, doctor, I never take medicine." I had no more sense than to laugh and called her attention to her inconsistency. Her answer was, "I'll have you know I'm boss around here." That, I was willing to admit.

One neighbor was a native of Kentucky. One day she told me this tale. She was raised in the mountainous area of the eastern part of the state, far off the beaten track. About the only strangers they saw were the "Yarb" (herb) buyers who came through. One night the "Yarb" buyer stayed at her father's cabin, and during the night became violently ill. They called the only help available, an old woman who was the local healer. After looking him over, the woman gave him three lead bullets and told him to down them. He was so sick he did it without question and immediately started improving. Then he began worrying about those bullets and what would become of them. When he asked the old woman she said, "Oh, they will be all right."

"Are you sure?"

"Sure I'm sure. Them bullets has been through at least a dozen men."

The grocery store was near where we lived when we first settled in the community, so I usually walked to get supplies. On the way I passed a house where the woman was usually outside, puttering with flowers or chickens. She was a big, friendly, jolly woman, much overweight, with a booming laugh and a ready joke. Frequently I'd stop to chat with her, though I never found out much about her and nothing about her theological ideas. One night I was called to see her in an attack of gallstone colic. A shot of morphine soon gave relief, and she started her usual kidding, centering on how easy I had made $5.00 just by giving her a shot. In the course of the conversation I asked, "Why didn't you try Christian Science on those gallstones and save money?"

Quick as a flash came the reply, "Young man, I'll have you know I have some sense with my Science."

I found out later she was a pillar in the local Christian Science organization.

Jack Dolan was a little man, short and skinny. He would weigh maybe 110 pounds if he was out in the rain after a big meal. But he was tough as leather. His wife was a big woman, tall, big frame and well upholstered. She was a jolly soul, always laughing and joking.

Another woman, rather elderly, was quite nosey, always interested in what others were doing and plain spoken in her comments. One day this woman stopped Mrs. Dolan on the street and said, "Mrs. Dolan, ain't your man right real well?"

"He was all right when he went to work this morning. Why?"

"Well, he is such a weeney, scrawny little man and you are such a big strong woman, I just thought maybe you were too much for him."

Ruby Dolan just loved to get Jack in a crowd and tell that story.

A neighbor of mine, a retired farmer, told me this.

A certain professional man, who, later attained a position of some prominence in the state, lived for a few of his younger years in Tillamook, and had been a friend of the above farmer. He was a big, powerful man, a football star in college, and interested in all outdoor sports. As the story was told:

"Me and Tom started out early one morning to go fishin'. It was a pretty chilly morning, so we had a couple of nips to warm us up before we started out. Tom stuck the bottle in his pocket, and I guess he was cold, for he kept nippin' at it as we fished. We worked up Penny Creek, not havin' much luck, until we came to Schmidt's land. In the pasture was Schmidt's bull. I knew that bull and I didn't want to have anything to do with him. He was just plain ornery mean. He had mauled two of Schmidt's hired men—guess he would have killed ’em if Schmidt hadn’t been handy each time with a pitchfork. I reeled in my line and told Tom I was going to walk around that field.

‘I wouldn’t walk around a field for any damned bull,’ answered Tom, and climbed over the fence.

“I started around the field, keeping an eye on Tom. When he was about a third of the way across, here come that bull, tail up, a-bellerin’ and a-snortin’. I thought Tom was a goner, sure. I was too far away to do any good; the only weapon I had was a pocket knife. The bull came up to about fifteen feet of Tom and stopped, pawing dirt and bellerin’ fit to kill. Tom kept right on walking and the bull turned and trotted off.

“When we got together again on the other side of the field, I said, ‘That was a close one.’ Tom just looked at me and said, ‘Huh, that bull knows who is the best man.’ ”

There was an Englishman living near Garibaldi; he had lived there many years when I first knew him. He was a tall, thin man, with a big moustache. He would answer pleasantly when addressed, seemed pleased to be noticed; otherwise he spoke to no one and never smiled or laughed. He always wore breeches and leggings. He owned no car, but took long walks about the country, walking with a long swinging stride, carying a stick. His bearing must have been the result of many years of military discipline. You could almost set your watch by the time he came to the post office. He lived alone, never worked, yet seemed to have all the money he needed. Gossip said he was a remittance man. Maybe he was.

One day I was called to his home. It was furnished as sparsely and plainly as a barracks, but everything was spotless. There were several pictures on the walls, mostly of soldiers, but in the central place was a family group in front of a large house that could have been in England. I wanted to ask a lot of questions, but did not think it polite. I am sure there was an interesting story in that situation, but did not have the nerve to dig for it. If he ever told it to anyone, I never heard of it. Maybe it is best that it died with him.

One thing I did have the nerve to ask about. There were two alarm clocks. Such an unusual situation seemed a harmless subject for a question, and I asked it.

“Oh,” he replied. “You see it’s this way. I set one clock to go off when I want to get up, and I set the other for thirty minutes earlier. When the first one sounds off, I just lie there and laugh at the bally thing." He was quiet a moment, then said, "And they say an Englishman has no sense of humor."

There was a fellow in our town who could not read or write, but he had managed to build up a teaming business. He had several horses, wagons, and other pieces of horse-drawn equipment. He would do anything that could be done with horses.

Plans were made to put up a business structure on a site that was on a steep hillside. Leveling off the site was quite a job; there was a lot of dirt to be moved, and as the slope was irregular, the yardage was hard to figure. The job was large enough that it interested contractors from quite a distance. Engineers spent days with transits and slide rules. John had not intended to bid on it; thought the job was too big for him. The advertisement also called for a performance bond. He had never had a bond and didn't know how to go about getting one. However, on the day the job was to be let, one of his friends who had sometimes worked with him, persuaded him to go look at it. As this fellow told it:

"John stood and looked at that hill for about five minutes, took a chew of Star, spit twice and said, 'There's about so many yards in that, and I can move her for so much a yard.' On the basis of those figures, I wrote up a bid on the back of an old envelope and we got the job. John was known to be so dependable they didn't ask for a bond. He missed the yardage by less than five yards, and our expenses were almost two cents a yard less than his estimate."

This same man switched to gasoline with the changing times. One year, while I was on the school board, he put in a bid on hauling school children. He was low, but there was the question of equipment. I knew he had no bus, and doubted if he had money enough to buy one, so asked him about it.

"Well, I know where I can buy a good used bus body cheap that will seat the number of kids specified, and I have a 'chastity' to put her on."

We gave him the contract and he performed satisfactorily.

People develop a remarkable indifference to a danger that is always present. Shortly after going to Tillamook County, I was called to see a family that lived near the fork of the Nehalem River.

The road comes close to the river in places. They told me where to stop, walk to the river bank and call. They would get me by boat. The little girl that came over to get me was too little to go to school, but she handled that boat like a fisherman. At that place the river is possibly a hundred feet across, quite deep, with a strong current. The mother told me that most of their contact with the world was across the river. Roads on their side were impassable in wet weather, which is much of the time in that country. I asked about school, and was told the school bus stopped where I had parked. The school kids had a boat of their own, rowed across in the morning, tied it up until evening when they rowed back. I asked if she wasn't afraid to have the little kids alone on that stretch of water, especially in the winter when it was dark by the time the school bus arrived, and maybe storming. It looked dangerous to me. She said she wasn't worried.

"I'd be afraid they would fall in," I said.

"Oh, they do fall in sometimes, but they always crawl out."

She raised eight kids on that river, and I never heard of one drowning.

This tale was told me by Oscar Tittle:

"When the Wilson River road was being built across the Coast Range, I worked one summer with a gang building culverts and small bridges. We worked well ahead of the grading gang, up in an area where there was nobody living. The Wilson River was wonderful fishing then, practically untouched, so I always took my gear along, and each noon, while the rest of the boys were sitting around eating, I'd put my lunch in my pocket, take my pole, and go down along the river. One day, while following up a little tributary creek, I found a patch of hazel brush, heavily loaded with very fine nuts. Hazel brush was rather rare in that country, and I resolved to have these nuts when they were ripe. The gang moved on to other culverts, but nearly every Sunday I would come back to look at those hazel nuts—*my* hazelnuts, as I got to calling them. I'd go around the patch, looking for any trace of anyone else that might have been around, but the only tracks beside my own that I ever saw were a set of bear tracks. As far as I could tell, no other human being came into that little hollow all summer, and the only animals I ever saw were a few squirrels. Never even saw a rabbit or deer sign. I was careful never to go in twice the same way, so as not to leave a trail that might arouse someone's curiosity.

"As summer ran out and fall started in, I watched the ripening of those nuts. At last I thought they would be ready to pick by the next Sunday. I estimated there would be about three or four bushels, maybe more. When the next Sunday came, I put some sacks in my old pickup, drove up the construction road to the point nearest *my hazelnuts,* and hiked in. When I got there, I could find only a few nuts, and not one was good. Careful examination showed nobody and no bear had been there since the last rain. Those squirrels had agreed with me that the nuts were ready to pick, and they had cleaned the patch, leaving only the culls. How they knew which nuts were no good was a mystery—I had to crack them to find out.

"At first I was pretty mad, and started back to the truck to get my gun. I'd teach those thieving squirrels to steal *my* nuts. But before I got to the truck, I thought maybe those nuts belonged to the squirrels and I was the thief. And they hadn't tried to kill me. I felt sort of small as I drove home."

Two young fellows were coming down the North Fork of the Nehalem River in a canoe. They came to where a small spruce, possibly a foot through, had blown down across the river. Where the current was strongest, the tree trunk was about a foot above the water. As they approached it, they decided that if both laid flat on the bottom of the canoe, they could just slip under and save some hard work. Just as they got to the tree, the man on the front of the canoe thought he had a better idea. He stood up, and at the proper moment jumped, expecting to come down in the canoe as it went under the log. However, the front of the canoe, relieved of his weight, bobbed up and lodged against the log. He came down in about eight feet of cold water. The canoe swung around; the stern, weighted down by the other man, slipped under the log, and the strength of the current pushed the canoe along until it was free, and was rapidly drifting downstream. The stern man, flat on his face in the canoe, was not aware of what had happened. The other man was swimming after the canoe as fast as he could, and, whenever he could get his mouth out of the water, yelling, "Wait, wait, wait!"

Another unique character was Miami Belle. She disappeared

from the scene before my time, but there were many tales about her. She lived some place in the woods on the hills above the Miami River, therefore the name. The story is that she made her spending money trapping, which was an unusual occupation for a woman to follow. It was said she and her old hound could follow a cougar track as well as any man in the county. And she was a dead shot. There were a number of men who did a little hunting and trapping, but she turned in as many pelts as anybody; mink, muskrat, fox, coyote, wolf, beaver, cougar, it was all the same to her.

She was not all there mentally, but never did anybody harm, and in her strange way tried to help people. She would appear at an isolated cabin with a gift, colored leaves, a bright colored stone, or sometimes a fish. If she saw someone fishing and having poor luck, she would show them where and how to do better. Or, if she saw a wagon stuck in a mudhole (and there were lots of them up that Miami River road), she would work hard to help them get out.

She had several delusions. One of them was that everybody was trying to cheat her. One day she was in the drugstore in Tillamook to buy something, probably some salve for minor injuries. The price was fifty cents, but she thought she ought to get it for forty-five. The druggist was stubborn; he wouldn't come down. She tried and tried, but finally gave up, paid the fifty cents, picked up her parcel and started out. By the door was a rack of writing paper. She picked up a five-cent tablet and walked out. The druggist saw it but said nothing. He knew he had lost the argument.

Another of her delusions was that she was beautiful, though she was as homely as a mud fence. Her features were coarse and irregular, and she had a large scar on one cheek, caused, it was said, by an argument with a cougar. Her hair was stringy, greasy and uncombed. She was tall, nearly six feet, with broad shoulders, long legs and big feet. She walked with a long swinging stride that would cover as much ground as any man could. She thought nothing of walking to Tillamook and back in a day. I do not know just how far that would be; it was about 10 miles from Tillamook to the mouth of the Miami, but how far up the river she had her cabin, I did not find out.

Her idea that she was beautiful probably influenced her choice of clothing. When out in the woods she wore overalls, a man's shirt and a black felt hat, but when she came to town she really duded up. She had several dresses; some brightly colored velvet, some of other materials with big, bright floral designs. All were dirty.

The striking thing about all the dresses was the hem line. Behind they just cleared the ground, which was customary at that time; in front, they were cut off a little above the knee (forerunner of the mini-skirt, I believe). This, of course, scandalized the women. At that time, nobody was supposed to see higher on a woman's legs than the ankle bones, and to wear trousers even in the privacy of home was terrible. It was a far cry from modern pants, so tight that they look like they must be put on with a spray gun.

I never talked to anybody that had seen her home, though all the oldtimers remembered her and had tales to tell of her strange actions. She was said to have a pet cat. Also it was said she raised a few pigs, somehow finding sustenance for them. Maybe she fed them the carcasses of the animals she trapped. If she did raise pigs, she must have eaten them herself, for nobody remembered her offering to sell a hog or any hog meat.

What was her end? I do not know. She just dropped out of the picture. Probably met with some sort of accident some place in the hills.

Lars Swenson was a big man, and he had enormous feet. His feet were so big he couldn't buy shoes in the stores; the factories just didn't make shoes that big. So, his shoes were custom-made by a shoemaker in Portland. One time he ordered a pair of shoes, and when he opened the box, there was only one shoe. Next day another box arrived with the other shoe. The story came out in the local paper, and the reason assigned was that the truck that brings the mail from Portland didn't have room for both shoes the same trip.

There was one character in Garibaldi who is worth a few moments' consideration. He was Joe Phelan, commonly known as Joe Whitewash. He seemed to be a little touched in the head, very peculiar in some ways, but pretty shrewd in others. Where he came from nobody knew; one spring day he was there. He slept under a big log on the beach at first, cooking his food over an open fire.

A beachcomber and his cabin.

He started gathering up odds and ends of lumber on the beach and around town, and before fall had accumulated enough to build a little shack in a grove of alders by a small creek that flowed into the Bay. He whitewashed the trunks of the alders, thereby getting the name Whitewash, by which name he was known for many years, until most people did not know his real name or how he got the nickname.

He made a little money by digging clams, catching a few fish, or any odd job he could get. He was a good worker, and rather ingenious at figuring out how to do unusual things. He would dig, make a lawn, plant a garden, build a fence, or do any other laborious job. His charges were ridiculously low, and if he thought people were hard up, he would work for nothing. He was scrupulously honest, insisting on paying for everything he got. Lots of the women he worked for would give him food. If they did it before he had his pay, he would reduce his charge; if after he was paid, he would do something else for them. Whenever one saw him about town, he was moving at a walk so fast it was almost a trot. One time he got a job longshoring, loading lumber on a boat. He was injured shortly after he started work, and when he was able to work again the other men refused to work with him; said he worked so fast he was a hazard to himself and to them. The fact that he had been hurt gave some color to the claim, but there was always a bit of question as to whether or not the real reason for the objection was that he showed up some that were rather "work-brickle," a local synonym for lazy.

He would eat anything, always investigated the garbage cans back of the stores, looking for things thrown out as not salable. When he bought food at the grocery store, it was always a half gallon of strawberry jam, a loaf of bread and a chunk of bologna. He would go down the street eating it, dipping up handfuls of the jam, tearing off chunks of the bread and sausage.

Sometimes he would buy a chunk of boiling beef at the butcher shop, which he would cook over an open fire, for he had no provision for a fire in his shack. Once he told the butcher that after he had cooked the last piece of meat he left it on the table to cool while he went out for some water. While he was gone, the rats got the meat, dragged it off the table and to a corner of his shack. As he told it, "I didn't get to eat a quarter of that beef."

He had made a practice of taking bones for his cats from the butcher shop waste barrel. A new meat cutter trimmed bones much more closely than his predecessor, leaving very little meat on them. After he had been there a few days, Joe came in boiling over, accusing the new meat cutter of trying to starve the cats.

One family had had a bathroom built on their house, and hired Joe to clean up after the carpenters and plumbers. They told him they had no use for the old privy; told him to tear it down and fill up the pit. Joe said if they did not want it any more, he knew a family that had none and would be glad to have it. They told him he could take it, so, when he was through that job, he got the privy on his shoulders, with his head through the hole (it was a one holer), and carried it along the main street to the family he was trying to help.

One woman regularly hired him to do her yard work. One spring he was cleaning up the winter's debris, digging flower beds, etc. He decided that yard needed a lily pool. He knew where there was a discarded bathtub, got it and brought it to the yard. Before he got it sunk into the ground, the woman discovered what he was doing and vetoed the idea. She didn't want a lily pool, and if she did want a pool, she would not want a bathtub pool, period. Joe was so mad he would never work for her again.

He had some peculiar ideas about clothes. Most of his clothes were cast-offs, but sometimes he would buy a pair of overalls, or a rough shirt. Underwear he would not wear, said it was weakening. Overalls and shirts were never washed; he wore them until they fell apart. When he bought a pair of shoes, he would soak them in the bay for three weeks. Said it prevented gangrene. Socks once put on were never taken off until they disintegrated. He said it was bad luck to take off socks.

He was a veritable pack rat. Anything he found that he thought could possibly be used, he would bring home. His shack was piled high all around the walls with all sorts of junk, so much so that two people could hardly get inside. I could not figure out how and where he slept.

With his ideas of personal cleanliness, it was inevitable that he would be lousy. He complained that he could not sleep because of the greybacks; he had to get up, take off his shirt and overalls and shake them to get a few hours rest. Somebody told him that if he would soak a rag in kerosene and put in on his stomach at night, he would keep the lice away. He tried it; said it didn't help much, besides he didn't like the smell of kerosene. Somebody else told him that if he would burn some sulphur in his shack it would kill the lice and the rats, too. He got a pound of sulphur at the drug store, put it in an old bucket and set fire to it. Then he went out and closed the door. Pretty soon the whole cabin burst into flame, and in a few minutes was just a pile of ashes. Probably the sulphur made a fire hot enough to burn a hole in the bucket. Anyhow, he killed the lice in the cabin; don't know about the rats, maybe they ran.

One of the boys around town asked Joe to sleep at his house until he had another place of his own. His basement had an outside entrance, and he had a camp cot and some old blankets that he was willing to burn if necessary. When Joe came, he took off his clothes in the street before he came in the house; said he didn't want to bring his lice into the house. He didn't either; they never saw a louse.

The fellows around town took up a collection and bought lumber for a new shack, and helped him put it up. Everything was new and clean to start, but soon it was just like the old one. That was the way he wanted it.

When the Second World War came along, Joe was much interested, talked about what beasts the Germans were, and how necessary it was to defeat them. Nobody found out where he got the idea; he must have had a bad experience with a German sometime in his wandering. He listened to the publicity on the radio at the tavern, and asked if he could buy a bond. They told him who was selling them in that district. He went to her house and told her he wanted to buy a bond; he did not know how much, but he wanted to use all the money he had. She asked him to come in and offered to count what he had. He walked up to her dining room table, with its clean white cover, and began to place on it folded and wadded up dirty bills, digging them out of various pockets, his hat, and his shoes, until there was quite a pile of them, and asked her to count it. There was a little over $1,500. When asked how much he wanted to use, he said all of it. She made out the necessary paper, and when it came to the name of co-owner he did not know what that meant. She explained it was who would get the money if he should die before it was repaid. He was much surprised; said he did not want it back, he was giving it to the government. He wanted to help lick those Germans. After a time, she was able to get it through his head that it was a loan, not a gift, and he gave the name of a sister. The tablecloth had to be washed after that sale, but its owner thought the experience was worth it. The fact that somebody wanted to *give* the government something sort of restored faith in humanity.

Joe lived to be an old man, but he never surrendered his independence. He lived in his own little shack, on food he paid for by his own labor. He never stopped trying to help; he would cut wood, or do other chores for the widows and the sick, refusing pay for it. He loved children and would stand for a long time watching them at play, with a most peculiar expression on his face, but he never spoke to one unless the child took the initiative. Most of them were afraid of him; their parents had warned them against him, and he seemed to realize it. He was always eager to work on any project he thought would benefit the kids, such as improving the school playground, or making better paths to the schoolhouse.

The big boys would break into his shack and steal or smash his worthless treasured junk, then hide in the nearby brush to hear him rave when he came home and discovered the vandalism. People who should have known better would torment him until he was furious. His tantrums of rage were thought to be funny. He would yell and tear around, but he never hurt anybody or damaged anyone's property.

When he was almost eighty-seven, he became ill, and the authorities took him to the hospital. He was a Catholic, though he had not worked at it much, and the priest sort of took charge. He got in touch with Joe's sister, and she came shortly before he died.

The sister was a nun, a very pleasant, gracious lady. She said Joe had been normal as a boy, and finished high school. When about 20, he was in an accident, in which he incurred a head injury. Following this injury, he began acting queerly, and finally wandered off. They had not heard from him for many years, long before the war—she was not sure just how long. Where he had been was never determined. His parents were dead, as were all the rest of the family.

He still had his bonds; they were found in his cabin. Also, he had accumulated some more money. He had enough to pay for all of his hospitalization and funeral expenses, and a little left over for his sister. Her vows prevented her from keeping it; I suppose she gave it to her order.

Who can tell what went on in that warped brain, or how strong was the temptation to lash out at those who teased and tormented him. Under handicaps that would have stopped most people, he did what he thought was right, treating people as he wanted to be treated, living the way he wanted to live, and leaving this world a little bit better for the many kindnesses he showed to those he thought needed them. I think he did the best he could with what he had to work with, and more cannot be expected of any man. Most of us do not measure up to that standard.

The Melting Pot, Etc.

The people of Tillamook County represented a wide variety of national and regional backgrounds. The larger groups were the Swiss and Scandinavians, but there were Russians, French, English, Italians, Mexicans and Slavic people that I knew, possibly others. Also, there were people from all parts of the United States and Canada. Among such a mixture, it was likely there would be some interesting individuals; and there were.

One was John Jugasek, a Yugoslav logger. At one time he had an injury that called for a rather long series of diathermy treatments. He was an interesting chap and I liked to talk to him whenever I had time. One day the conversation turned to hunting and I asked about hunting in Yugoslavia. He told me he had often hunted in the mountains on the line between Yugoslavia and Albania. When asked what game was to be found there, he replied, "I shoot deer, rabbits, squirrels, quail, and sometimes I shoot an Albanian." He paused a minute, then added, "Sometimes Albanian, he shoot first. Then I run like Hell."

Old Joe had a little tailor shop in Tillamook. He was a Russian, a refugee from the Russian revolution. I do not know how he found his way to Tillamook, but there he was. Shortly after the start of World War II, I asked him to make me a suit, but he said, "I no got cloth. Can't get."

As time went on, I asked the same question at intervals and received the same answer. I was eventually down to one pair of pants, and that pair was getting very, very thin in certain essential spots. I was contemplating the necessity of going about in overalls. One day I met Joe on the street and as a matter of routine, but without hope, asked, "Joe, you got cloth yet?"

Much to my surprise he answered, "Ja, I got cloth."

"O.K. Make me a suit."

I knew he had my measurements and I was beyond caring about little things like texture, color and pattern. I just wanted a pair of pants I could trust. But to my surprise he said, "No, I no make you suit." Wondering if, through language difficulties I had offended him in some way, I said, "But why, Joe? I need a suit."

"My cloth, she no good. I make you suit, every time you see that suit, you think, 'Huh, Old Joe he make me that suit. Old Joe, he hell of a guy.' No I no make you suit." And he didn't, in spite of all I could say.

Fortunately for me, shortly thereafter he managed to get his hands on some cloth he thought suitable.

Seems to me there is an editorial, or a sales talk, or something in that little story.

Several years ago, the county decided to build a new bridge across East Beaver Creek. It was to be of reinforced concrete. The excavating was done, and the forms built. Then two men were detailed to put in the reinforcing. This was the usual half-inch steel rods, placed criss-cross, and wired at all the intersections. About ten o'clock, one of the men started complaining of a toothache, and by noon it was hurting so much he decided to take the afternoon off, go to Tillamook and see a dentist. The other fellow didn't want to lose the time, so it was agreed that the toothachy one should take the truck, but be back by quitting time to pick up his partner. The man who stayed on the job told what happened this way:

"I was working along all by myself, and it was kind of lonesome. This bridge was to carry a new cutoff road, so there was no traffic. As a matter of fact, nobody but our crew had been around for the last week. The bridge was not a large one, and working room was limited. For a part of each section, I had to lie with my head through the grid, to reach the last places to be wired. I had just finished one section, and since I couldn't reach my belt holster, I tossed my tools to where I planned to start the next section. Then I tried to get my head out of that grill, but somewhere I'd made a mistake and had wired some rods so close together that my head wouldn't come out. At first I was amused that I would do such a stupid stunt. But when I tried to untwist some of the wires to let the rods shift, I couldn't do it; I'd done too good a job of wiring those joints. My fingers couldn't loosen the wires and my tools were out of reach. There was not one joint I could loosen. No matter how I twisted and turned, my head wouldn't come out. I was caught like a cow in a stanchion. I worked at those joints until my fingers were bleeding. I tried every way to get my head loose until I had worked skin off my jaws and the back of my neck. No soap! I wasn't too uncomfortable; it was a warm day and I had a smooth board of the form under me, but the flies did bother my cut places. There was nothing else for it but to lie there and wait for my partner to come back.

"When he came and saw me, he started to laugh. In spite of the fact that his jaw was swelled so his head was lop-sided, he laughed and laughed. By that time I couldn't see anything funny in the situation, and his laughing didn't help any. After a time he stopped laughing long enough to say, 'I thought all the beaver were trapped out of this creek a long time ago.' Then he laughed some more. When he could stop, he said, 'I believe I'll get the rest of the crew.' (They were working on another bridge something over a mile away.) And he started towards the truck. But he couldn't make it; he had to stop and laugh some more. At last he relented, picked up my tools, cut two or three ties, allowing one rod to shift a little, and I was loose.

"By the time we got to the county shed, all the rest of the crews had reported in and gone home. The only one around was the night mechanic, and he was deaf, so my partner didn't try to tell him the story. By morning I could see the humor of the situation and didn't much mind the kidding I got."

A rancher told me this tale.

"I like frog legs. There are a lot of frogs in the slough back of my place, big ones. Sometimes I'll go out there at night with a flashlight and a twenty-two rifle. As they sit on the bank, their eyes reflect the light, and a bullet just above the eyes will knock them over. A frog is hard to kill. The bullet stuns them for a while, but they will wake up and kick around for several hours, even with the top of their heads shot off. Usually I'd get enough for a mess for the family, take them home and skin out the legs before I went to bed.

"One day the fellow that ran the hotel dining room offered me a good price for frog legs. I needed the money, so that night I shot all the frogs I could put in a gunny sack. It was sort of late by the time I got home, and I was tired, so I threw that sack of frogs in the bathtub, with the idea of skinning out the legs next morning, as soon as I finished milking.

"During the night the baby started to cry and my wife got up to see about him. She didn't bother to light a lamp, and was barefooted. Pretty soon she screamed louder than the baby. I jumped up, made a light, and soon found the trouble. That sack full of frogs had waked up; most of them had gotten out of the sack, and a lot of them had jumped out of the bathtub and were hopping around the house. She had stepped on one.

"Maybe you don't think that woman was mad. She wouldn't speak to me for three days, and made me sleep on the sofa. She never would cook frog legs for me after that; if I wanted frog legs I had to fry them myself, and she wouldn't touch them after they were cooked, though she used to like them. She wouldn't even wash the dishes the frog legs had been on; she would put them out on the back porch, and there they set until I washed them.

"For some reason, frog legs didn't taste so good after that."

One day a bunch of fellows were hanging around the pool hall in Tillamook. A stranger came in, went up to the bar and ordered a glass of beer. He kept his right hand in his side coat pocket, and there was a sound like a baby chicken chirping. When his curiosity got the best of him, one of the hangers-on asked the stranger about the sound. The newcomer seemed like a friendly fellow; told them he had a four legged chick in his pocket, wrapped in a handkerchief to keep it warm, and he kept his hand in his pocket to protect it, just in case something bumped him. He expected to make a lot of money showing it when it grew up, so was taking good care of it. Someone asked if they might see it, but the stranger was a bit doubtful; he didn't want to take a chance on dropping it and hurting it. The boys all promised not to touch it, but they did want to see such an unusual sight. It was finally agreed that if the group would hold their hands around the edge of a table for a fence, so the chick could not possibly fall off, they could see it.

They cleared a card table; everybody got around, making a fence with their hands. The stranger pulled out a wadded up handkerchief and carefully placed it on the table. The cheeping kept up. The stranger observed, "Baby is getting hungry. I'll just feed her while she is unwrapped."

That pleased everybody. They could get a longer look. The stranger carefully unwrapped the handkerchief and revealed—nothing!

He was a good ventriloquist.

One man of the community had lost his arm just above the elbow. It happened when he was a child, and he had made a good adjustment. He could do nearly everything with his left hand. He was an intelligent, industrious individual and had made a place for himself in the community. He had quite a sense of humor, and could laugh at jokes about the arm.

One day he was in a neighboring community. As he was talking to a friend, a small group of little children gathered around, staring at a man with only one arm. They had never seen such a thing before. One little boy mustered enough courage to speak and said, "Mister, how did you lose your arm?" The man looked down at the stump, pretended surprise, and said to the boy, "Why, I have lost it, haven't I? Now where do you suppose I could have done that?"

Then he pretended to look all around and said to the boy, "I don't see it. Will you help me look for it?"

"Sure. Come on kids, let's help the man look for his arm." And the whole bunch scattered, each one looking for the missing arm.

The same chap told me this tale:

"I had been out to Portland and was coming home rather late. There was a moon but it was so cloudy it was quite dark. Along the Sourgrass road I saw a young fellow walking and asked him to ride. He got in and I drove on. I noticed that he sat just as far from me as possible, had nothing to say, and kept watching me. Finally he said, 'Mister, if you want to, you can stop and search me.'

"Why would I want to do that?"

'You are holding a gun on me.'

"No I'm not. What made you think that."

"Just then the moon came out from behind a cloud and we could see one another fairly well.

'Why you only have one arm,' said the fellow in surprise. 'You were driving with one hand and holding your right arm close to your side. I figured the only reason was you had your right hand with a gun down on the seat where it was dark. I'm a hitchhiker, and a stranger, and I thought you were not taking any chances.' "

I do not know why it was so, but around Tillamook County

A typical home in the backwoods.

we seemed to have quite a number of single, old men. Many of them probably had never married; many old-time loggers didn't. Some of them lived with, or near, friends or relatives. While some of these became a bit crotchety and difficult, most of them preserved a more or less normal relationship to the world. Those that elected to live away from other people frequently developed peculiar characteristics. I wasn't sure whether they became queer because they lived isolated lives, or chose to live isolated lives because they were queer to begin with; probably some of both.

Most of these loners got far back in the hills where it was not too difficult to find a secluded spot, near water, where they could locate a cabin. If they were not careless with fire, the timber companies didn't make much objection if it was ever found out that they were there.

Their cabins were usually very primitive affairs, just a one-room shelter, made of odds and ends, either lumber picked up here and there and packed in on their backs, but more often of poles covered with hand split shakes, riven out on the spot. Sometimes there was a plank floor; more often just hard packed clay, built up a few inches above the surrounding level. They would have been hard to keep clean, if one had tried, and nobody tried.

How these old codgers lived was rather a mystery. It was before the time of our present elaborate welfare system, though a few of them did get a small allowance from the county. Some of them would occasionally take on a little job, like helping a farmer with his hay, or cutting a little firewood. Some did a little trapping, or peeled chittim (Cascara bark), or dug medicinal roots, or collected fir pitch. Some even bootlegged a little venison. They didn't need much cash money. The streams were full of fish, deer were fairly plentiful; a fellow who knew how could find edible plants, roots or berries, and a small clearing in a fertile spot would grow all the spuds and tobacco they needed without the expenditure of too much labor. (The tobacco was pretty stout, but that was the way they liked it.) Their clothes were usually somebody's cast-offs, never washed, worn until they literally fell apart. Footgear was usually cast-offs, frequently not mates, or crude hand-made moccasins of deerskin.

No matter how isolated a cabin might be, somebody knew of it, and if they didn't see the old fellow who occupied it go up the trail for a few days, they would check up. One old chap that was reported missing, lived somewhere back of the little town of Netarts. Nobody knew just where his cabin was, but some of the people knew where the trail to it started; they had seen the old coot start up it, usually with a small pack of supplies on his back. The report that he had not been seen for several days was made to the county nurse. She was an energetic, outdoors-type gal, and the weather was nice, so she decided to hike back up that trail and find the cabin. She found it without trouble, and in it, on a pile of old sacks, tattered blankets and deerskins which passed for a bed, was the old fellow, dead, cause unknown. Evidently he had had some prostatic trouble—there were catheters all over the place, and each catheter was connected to a tube leading to a crack or knothole. There was no place in that cabin where one could be more than arm's length away from a catheter connected to the outside. Talk of conveniences—he had 'em!

Old Pete had a cabin on the beach. He did odd jobs for a little money to buy a few necessities, but mostly he lived off of handouts, fish and clams he could rustle from the ocean, even garbage. Once he was violently sick after eating some beans somebody had discarded. "I thought they tasted a little sour, but not too bad," he said.

One day he fell off a rock on the beach and fractured a fibula, the small bone in the lower leg. This is a painful accident, but requires little treatment except rest. Since I knew where and how he lived, I took him to the hospital to be off his feet for a few days, and arranged for the county to pay for his keep. A few hours after I got back home, the hospital called, saying Old Pete had disappeared. They had looked about the building, then called the police, but no trace of him was found. I looked around some the next morning and found him in his cabin. When I asked why he did not stay in the hospital he said, "Me lay in bed and let a little nursie girl wait on me? No, no!"

When I asked how he got home, he said he walked along the railroad track; afraid if he walked on the road someone would see him and take him back to the hospital. He had picked up a broken piece of a plank which he had used as a crutch. Twelve miles hopping along a railroad track with a broken bone in his leg—quite a night's hike!

Another old fellow had been a lifelong deer hunter. He lived quite a piece up in the woods, but what was very unusual, he had a road up to his cabin, passable with a pickup in dry weather. A nephew lived a mile or so from his cabin, and kept on eye on the old man. One summer this nephew brought him to the office several times. His heart was failing rapidly. His last visit was just before the deer season. After the usual routine he said, "Doc, I want to go hunting tomorrow."

"That would be foolish, Sam."

"But I want to shoot one more buck before I cash in my chips."

"You never could stand it to hike up in the hills."

"There is a place near my cabin where you can nearly always find a deer. I would not have to walk very far, and it ain't very steep."

"Don't think it is a good idea."

"Doc, I'd be willing to die to get just one more buck. I can't live long anyway, please let me try for just one more."

"Sam, you are old enough to make your own decisions. I'm your doctor, not your boss. I'll advise you, but I won't tell you."

Next afternoon, the nephew drove up to the old man's cabin. There was nobody home, so he started up the trail that had been his uncle's route into the hills for many years. About a quarter of a mile up, he found his uncle's body, his rifle by his side, an empty cartridge in the barrel. A short distance away was a trail of blood that lead to a dead buck, with just one bullet hole in him.

I think old Sam died willingly and happy. He killed himself just as surely as if he had put that bullet through his own head, but who is wise enough to criticize him?

My dad sometimes rode with me on country calls. One day we stopped where I knew I would have to spend some time, so Dad decided to look around a bit, for he had never been in that neighborhood before. He found a trail leading up into the woods and followed it.

When I was ready to go and we were driving off, I asked him where he had been. He replied, "Do you know Bill Slichter that lives at the end of that trail that takes off just east of where you parked?" I had heard of the old bachelor who lived in a cabin up there, an old time trapper I think he was, but had never seen him. "I went up to his place and visited with him for a while. He says he has been there about 20 years, but I don't believe him."

"Why not?"

"He has been there longer than that."

"What makes you think so?"

"Nobody could get that dirty in just twenty years."

Some of these old fellows kept reasonably clean, but some didn't care. As a result, the dirty ones were often lousy. One day I happened to see one of these old fellows brought into the hospital by the sheriff's men. He had been sick in his isolated shack for several days before someone found him and called the sheriff. As he was brought in on a stretcher, I thought there was something peculiar about his hair. That long gray thatch seemed to shimmer. Closer inspection showed that the hair was indeed moving with a gentle, undulatory motion. Mr. and Mrs. Greyback, their parents, their kids, their uncles, aunts and cousins, down to the umpteenth generation, were in that hair in force. I didn't look under the blankets, but am sure that more of them were there.

The man was put in a bathtub, his clothes peeled off, wrapped in papers and burned, then clippers were used to remove every bit of hair from his head and body. Then he was scrubbed with a brush and the outer layer of dirt removed, even if he was a sick man. After that the tub was scrubbed with Lysol!

Ed Pinkus was another one of those old codgers. His cabin was up Coal Creek. He had found a little cove along the creek where the ground was near enough level to be cultivated, and where he was protected from the cold ocean breezes. There he had slashed out a little clearing big enough to raise all the corn he needed for a year's supply of moonshine. He reckoned his crop in gallons, not bushels.

One year somebody told him pumpkins could be raised between the rows of corn. He considered the idea carefully, stimulating his thinking apparatus with nips of his previous year's crop. The more he thought about it, the better the idea seemed. It was double use for his land. Then he hit on the idea that if pumpkins would grow in the shade of a cornfield, they would grow in the shade of alder trees. So, when spring came, he bought a lot of pumpkin seeds, planted them with the corn, and over several acres of alder growth adjoining his clearing.

The season must have been good for pumpkins, for he had a wonderful crop; nice big pumpkins, so thick on the ground he could hardly harvest his corn, and all around under those alder trees. Then he began to wonder what to do with them. He had been so interested in raising pumpkins that he had never given a thought to what to do with them after he had them. There was no way to get a wagon up to his place, and there were too many to carry out on his back. Pumpkins are heavy and awkward to handle.

Finally he took a sample to the storekeeper, who was interested. The merchant had never heard of pumpkins being raised in Tillamook County. He figured if he put a pile of them in front of the store, every man around would start to thinking of the pumpkin pies his mother used to make. He made an offer to Ed, reasonable, if a way could be found to deliver.

The storekeeper's next move was to contact Hans Schneiter,

a "greenie," a recent emigrant from Germany, who had settled on a cutover tract on Foley Creek. He was doing general hauling to keep his family eating while he was getting his land cleared. The merchant told him he had bought some pumpkins, maybe six or eight wagon loads, from Ed Pinkus on Coal Creek. He assured Hans the farm was not over six miles from the store, but neglected to tell him there was no road to it. He offered about three times what Hans would ordinarily charge for a six-mile haul, but added that the pumpkins were perishable, and if not hauled promptly would spoil. He wanted whoever got the job to put up a performance bond to insure that the pumpkins would be at his store by a certain date. Hans was so anxious to sew up what looked like a plushy job, before his competitors heard of it, that he gave a mortgage on his wagon and team for the bond.

The storekeeper was jubilant. He figured he wouldn't have to pay Ed if the pumpkins were not delivered, and he would get a good team and wagon because they were not delivered. He was certain that the wagon could not get to Ed's place, and equally certain Hans could not carry that much pumpkin down to the road in the time available.

Hans soon found he had been suckered, but did not give up. After much thought, he put a pumpkin in the creek to see if it would float. It did, just barely, but it floated. So Hans went down to where the creek crossed the county road, near the Tilden home, and found a place where he could get a wagon to the side of the water. There he fastened a floating log diagonally across the creek so any floating object would be forced to that side of the creek. He left his boys with the wagon, while he went up to Ed's place. He carried all those pumpkins to the creek and set them afloat. When they floated down to the barrier and were crowded to the bank, the boys picked them out of the water. When Hans had all the pumpkins in the water, he followed down the creek to refloat any that had grounded on sand bars. Hauling those pumpkins to the store was simple.

Everybody was happy but the merchant, and he didn't come out too badly. Ed got paid for his pumpkins; Hans got back the mortgage on his outfit and a reasonable price for his labor and the hauling job. Many families had pumpkin pie, the first of their lives for most of the kids, and jack-o-lanterns began to appear for the first time in that country.

But there was a little static later. A bunch of the boys got together with a story, and each one told Ed that pumpkins make good moonshine, and if he had sold that crop in fruit jars, he could have gotten several times as much for it. After three or four of them told the same story, Ed believed it, and was he mad! Resolved to manage differently next year. But, unfortunately, he died that winter. The details nobody ever knew; some hunters found his body in his cabin after he was a long time dead. It was commonly supposed there was some connection between his principal product and his death.

I don't know whether pumpkins will make moonshine. Don't see why not; there is considerable carbohydrate in a pumpkin and it should be fermentable. Probably would not taste like Scotch, but I have been told frequently that customers for mountain dew are more interested in potency than flavor.

Earlier view of north jetty at Barview, entrance to Tillamook Bay, showing railroad built to haul rock. A similar jetty is being extended now on south side. Purpose is to confine water to narrow channel to prevent silting, a major obstruction to navigation. Small structure at end of jetty is Coast Guard lookout station. (Courtesy G. B. Abdill)

World War II Tales

When World War II came along, there was much concern, not only that shared by every American, but also because of the additional hazard of being in a possible invasion area. As we all knew, our whole west coast was wide open, and Tillamook Bay looked like a logical landing area. There were miles of good landing beaches. It was a considerable distance to any sizable military installation. There were good roads to north, south and east. All sorts of schemes were advanced, but a local volunteer defense organization quickly developed, which brought most of the plans under one head. It was promoted and headed up by a blind veteran of World War I named Stewart Arnold. The organization was rather loose, covering a lot of projects, but was fired with enthusiasm, determination and practical judgment. This Tillamook organization was the subject of a feature article in *Life* magazine. I do not remember all the angles of the organization; probably did not know most of them, but did have a slight connection with a few.

In a land of deer hunters, a large part of the men, and many women, not only had high-powered rifles, but could use them effectively, so neighborhood rifle squads sprang up everywhere. One branch of these was made up of men unable to walk much, due to age, accident, etc. Rifle pits were placed in strategic spots on the hills over possible landing sites and along the roads; a crippled man was assigned to each, and an able-bodied man or woman was designated to help him get there, with three days rations and water. No provision was made for getting him out. They figured if it was a successful landing, there would be no possibility of getting out. If the landing was not successful, there would be plenty of time for his buddy to come back after him.

One time a rumor was spread in the inland areas, around McMinnville, Forest Grove and Hillsboro, that the Japanese had landed on the Tillamook coast. Men grabbed their deer rifles, all ammunition available, jumped in their cars and headed for

Tillamook. Literally hundreds of armed men poured into that little town, much to our surprise, for we hadn't heard the rumor. It was a false alarm, but the spontaneous response made one proud of his fellow Americans.

Another line of preparation was followed by the dairy farmers. Recognizing that the stock of high grade dairy cows, some of the best producing stock in the nation, was important to the future of our country, plans were made to get some of them out. The military had informed us that in the case of invasion, all roads would be cleared of civilian traffic, freeing them for troop movements into the area. The civilian population could not be evacuated by road, much less animals. So, men familiar with the woods mapped out trails by which livestock might possibly be walked out. Old logging roads, deer trails, hunter and trapper trails were utilized to outline several routes that zigzagged through the mountains, finally ending on the other side. Only the most active of men and animals could possibly travel these routes. Assembly areas were designated and trail crews organized. It would be rough going for pampered dairy stock. Cattle that for many generations had done nothing more strenuous than walk from barn to pasture; cows that were heavy with calf, or heavy milkers, with large, heavy udders; cows that had been fed regularly and liberally all their lives, were in for a severe test. To travel all day, for several days in a row, over a trail that was fit only for a deer, steep, narrow paths, over rocks and fallen trees, with no grain and very little grass, only browse which cows will eat only as a last resort, would be a real trial. It would take a heavy toll, but probably some of them could make it, enough to preserve the bloodlines for the time of recovery.

Of course, the main burden of defense rested on the military, and what their plans included was their secret. We could see some of their preparations, such as armed men everywhere, mined bridges, buried charges of explosives so placed that roads would be closed, machine gun nests, patrol of the beaches day and night, often with dogs, the black-out, prohibiting any light from any source where it could possibly be seen from the ocean, even if only reflected from the clouds.

Many of our young people went into the armed services and some of them did not come back. Many old people assumed duties

Naval Air Station (blimp base) built near Tillamook during World War II (author's collection).

beyond their strength. Hundreds of new people came into the county for the military construction jobs, crowding every available house and eating place, increasing the demand for scarce consumer goods. Several thousand people took first aid training; first aid supplies were assembled; standby people were trained to operate necessary utilities and services; large amounts of blood were contributed to Red Cross; Red Cross Canteen and Motor Corps groups were organized; rubber drives, scrap metal drives, paper drives, bond drives were carried on; many women took jobs they had never dreamed of tackling; rationing introduced complications into every phase of life. But in spite of all that, life went on. The logging camps and sawmills operated, farmers milked cows, cheese makers made cheese, people ate and slept, women had babies.

Of course, our experiences were nothing compared to some others. We had no real hardships. But the steady way the rank and file of people faced the possibility of disaster, and met the little day by day decisions and adjustments, engendered a new respect for people. The way old people came out of retirement and assumed the duties vacated by those going into service was astonishing. The various projects threw all sorts together, mixing up classes and cliques. When you get to really know them, from working with them on a project to which all are giving their best, even if you don't like part of them, you have to admit that most people are mighty fine folks. We often think some people are made of pretty poor protoplasm, but maybe if we were in their place, we would not do any better than they do, possibly not as well.

Sometime during the war a military plane crashed against Point Lookout. Some men were killed, five as I recall. A sergeant came to make the necessary reports. I was County Health Officer at the time, and he came to my office to check on death certificates. He said he would need to find a quiet spot to work, a typewriter to make out the necessary forms. I offered a typewriter and a room. He worked steadily for nearly two days. When he was through, he came into my examining room to thank me for the use of office and machine, and we chatted for a few minutes. I made some remark about the amount of paper work it took to wind up the records of five men. He agreed, said at first it used to irritate him, but now he took a lot of satisfaction from it. I didn't understand, so he explained, "I know the United States

Interior of blimp hangar. Note size of door, lower left.

army can never be wiped out. There just isn't enough paper in the world to get them all buried."

During the war time, suspicious noises were heard on Mt. Hebo. Every day, at about the same time, and lasting for only a few minutes, there was a noise like a small gasoline motor. Mt. Hebo is heavily wooded, with nobody living on it except a government installation on the top—weather, fire prevention, etc. This noise was heard distinctly by people around the foot of the mountain, but they could not agree on where it seemed to come from. The general idea was that there was an illegal broadcasting station some place up there, sending messages to the enemy. Military and civilian agencies searched the area repeatedly, but without finding anything. The noise did not continue long enough at a time to allow a searcher to follow it. That is a big area, rough ground, heavy undergrowth, very difficult place to conduct a search, but repeated efforts were made.

One day a state policeman, recently transferred in from the eastern part of the state, was participating in the search for the hidden broadcaster, when the noise started. "That is no motor;

that is a grouse drumming," was his immediate reaction.

The grouse is not a native of that part of the state. None had been seen or heard within a hundred miles. How this bird got so far from home was never determined. Maybe he was blown in by a storm; maybe he was just a natural-born tourist.

Early in the thirties, the cultivated oyster industry was introduced to Tillamook Bay. Several people were interested in it, and as the industry originated in Japan, it was thought that Japanese workers would know more about tending the beds. So three or four of them were brought in, I don't know from where. They lived very quietly near their work, on the Bay Ocean road. They came to town only for supplies and bothered no one.

When the war came along, there was some talk about them, but nothing was done until all the Japanese were sent to a concentration camp. Before that occurred, a group at a beer joint started talking about these people. The more they drank, the more patriotic they became. The end result was that they decided it was their duty to run those — — Japs out of the country. A group, several times as big as the group of Japanese, went home to get guns, got in cars, drove out to the place, formed up in a straggly line, and shooting as they went, marched on the house where the enemy lived. The Japanese had no guns, but they had some firecrackers, and they set them off. The brave defenders of the American way of life thought they were meeting overwhelming resistance, so retired rapidly. One man was running so fast he didn't see a log in his path, stumbled over it and broke his leg. He walked with a limp the rest of his life. What probably hurt more than the accident was the sympathy he did not get because he was never awarded a Purple Heart. It was rank injustice.

During World War II there were troops stationed all along the coast. One contingent that was near Tillamook had a group from the slums of an eastern city, and they thought they were pretty tough, Local people, they thought, were real hicks. When on leave, they patronized the local taverns and swaggered around town, making themselves quite obnoxious. One night a group raised a ruckus in a pool hall and about a dozen of them landed in jail. To show their contempt for local authority, they proceeded to wreck the jail; broke out the windows, tore up the bedding, smashed the furniture, and even tore loose the wash basins and toilets. A state policeman went into the bull pen, unarmed, to remonstrate with them, and they ganged up to show him how things were done in the big towns. They used pieces of the broken furniture as clubs and went after him. What they didn't know was that Tom Butler had once been heavyweight champion of the Pacific Fleet. While he was getting on in years, had put on some fat here and there, was a little stooped and did not look too formidable, he was still pretty much of a man.

By the time I got to the calaboose, they were the sickest looking bunch of punks you can imagine. Most of them were stretched full length on the floor, along the walls of the bull pen, as far away from trouble as they could get. We hospitalized a few of them that night, one with a skull fracture, a couple of broken arms and others with less serious injuries. Those with just bruises, minor cuts, bloody noses, broken teeth, etc., spent the rest of the night on the floor of the bull pen they had wrecked.

Next morning, the colonel in charge of the area, a long-time Army man, came in, checked up on what had happened, called ambulances and sent the hospitalized boys to an army hospital. Those in jail were released to him, and all able to walk were returned to full duty with orders, "No favors." I happened to run across him in the hospital and remarked, "Quite a party the boys had for themselves last night."

"I hope they enjoyed it," he replied grimly, "for they will be paying for it for a long, long time. Not one of those boys will get another leave until this mess is paid for."

We never saw those boys in town again. Maybe the colonel meant what he said.

During the war, a jovial Swiss farmer was in charge of war bond sales in his part of the Nehalem Valley. Shortly after the start of one drive, I saw him and asked how he was getting along selling bonds. His face lighted up with a broad grin as he reported, "All done—a hundred percent sales and three times our quota."

I congratulated him and asked how he did it. He said, "Tom Tobec my partner. The first day of the drive, soon as I got through milking, I go across the road to Tom's house and say, 'Tom, let's go sell bonds.'

"Tom say, 'O.K., but let's have a little drink first.' So we did. Then I say, 'Tom, let's go to my house and have another little drink before we start.' So we did. Then we go to Anton Pavlick's and say, 'You want to buy a bond?' Anton say, 'Yes, I buy bond, twenty-five dollar.'

"I fill out the paper, but instead of twenty-five dollar I write two hundred and fifty, hold out paper to Anton and say, 'You sign.' So he sign. Then I look at the paper again, laugh, slap my leg and say, 'Ho, ho, ho! That good joke on me. You say twenty-five dollar and I write two hundred fifty. Ho, ho, ho! Good joke on me!' Anton, he get mad; he cuss like logger; say he won't pay, but I just say, 'You sign, you gotta pay.' Anton he got lotsa money, but he stingy. Two hundred and fifty he never miss. After a while he cool off and say, 'O.K. I pay. Have a little drink,' So we did.

"Then we go to Ferdinand Schmitz. He got German name but he not German, he Swiss. He got good farm all paid for, and sixty, seventy good cows. I say to him, 'You German, you got good farm in this country, you better help good.' Ferd say, 'I no German, but I help. I buy one hundred dollar.' I write out paper but I write five hundred dollar, and say, 'You sign,' and he sign. Then I look again, slap my leg and say, 'Ha, ha, ha, ha! That good joke on me. You say one hundred, I write five hundred. Ha, ha, ha, ha!' Ferd get mad and cuss and say he no pay, but I just laugh and say, 'You sign, you gotta pay. You German and if you no pay what you sign, I tell F.B.I.' After a while he say, 'O.K. I pay. Have a little drink.' So we did.

"Then we go to Sam Solomon's house. Sam's father leave his farm free and clear, but Sam no work it very good. Sam like to hunt too much and lots of times he late with milking. We say, 'You want to buy bond?' and he say, 'Yes, twenty-five dollar.' I write two hundred and he sign. Then I tell him I make mistake and he yell and cuss. He pretty good cusser. Then I say, 'I hear game warden check list of bond buyers' and he stop yelling quick. He say, 'You no say nothing, I pay. Have a little drink,' and so we did. This morning I find nice piece venison in my milk house. Don't know how it get there cause I kill no deer since last hunting season.

"Mike Kasek kinda old man, come to this country when he young feller. Got good farm all paid for and money in the bank. He like schnapps; has own still out in woods. Makes good stuff, too. I think his fadder taught him. Old Mike used to work in distillery in old country. Mike don't make much, just for self and few neighbors, maybe, three hundred gallons a year. He lucky, never get caught, but he always fraid. So he say he buy twenty-five dollar and I write two hundred fifty and he sign. When I say mistake, he say he no pay. So I say, 'That all right, just send along few jugs schnapps.' So he say right quick, 'I just joking, I pay. Have a little drink,' and so we did.

"We pull same trick on Louie Lietz and Tony Sullivan and Sam Trethic and Bob Thomas and Gil Broadus and all the other fellers with good farms all paid for. Not one of 'em read the paper before they sign. All got mad but all cooled off and everybody gave us a little drink to show we still friends. We sell lots of bonds and we feel good.

"But when we come to fellers not so well off, we just ask them to buy twenty-five dollar bond. Like Ole Hanson, that lost most of his cows in Bangs' test last year. Or Johnny Jenkins. His wife been in hospital more'n six months. Got cancer, and she gonna die. Or Pegleg Thomas. His bull got him down last year and nearly killed him. He had to hire a man to milk his cows for him for more'n' six months, and he didn't get a good man, either. Pretty poor farmer anyhow. Or Buddy Perkins. He young feller, just started farming for himself last year and he in debt over his ears. He hard worker and got good wife so he will make it O.K., but right now its pretty tough tittie. Fellers like that can't afford much, but all buy little bond. All glad to help, and nearly everybody say, 'Have a little drink.' And so we did.

"We see everybody in our territory that day before milking time. Then we go home. When we stop in front of Tom's house, I say, 'Tom, how much you buy?' and Tom say, 'One hundred dollars.' I say, 'Nothing doing, you buy five hundred.' Tom say, 'O.K. Come in and have a little drink.' So we did. Then Tom say, 'Joe, you make me buy five hundred dollar bond, you gotta do it too.' So I say, 'O.K. Come over to my house and have a little drink.' So we did.

"Then Tom say, 'Let's go to Wheeler and turn in our papers.' I say, 'O.K., but first let's have another little drink.' So we did. We feel good. Then we go to town and we first ones to check in. Nobody turned in as much over quota as we did. Me and Tom Tobec, we best damn bond salesmen in Tillamook County."

The Doctor's Day

Office hours for a country doctor are a never failing source of interest. You are always curious about what problem the next patient will bring in. I don't think there is any place in the world where so many amusing situations come up, and I don't think anybody has as much fun as a country doctor. The most interesting things in the world are people, and the country doctor gets them in large doses, in concentrated form and in great variety.

Art Linkletter says kids do and say the darnedest things, but I think it should be enlarged to include adults. The more you see of people, the more you have to respect them. You may not agree with them; you may think they have drawn the wrong conclusions; they may have acted hastily or imprudently, but when you can see things from their angle, you can't condemn. The country doctor has an unusual opportunity to get behind the scenes in many situations, to observe the abilities and responses of people.

Oregon early passed a law calling for premarital examinations. One day there showed up at the office a smiling, roly-poly girl of about 16, quite, quite pregnant. She declared she wanted an examination to get a marriage license. We went through the usual formalities, then chatted a bit, and I asked her how she felt about getting married under such circumstances. She was puzzled for a minute, then her face brightened; she rubbed her little round tummy, and said, "You mean this? Oh, I like it fine. We have been wanting to get married for over a year but our folks wouldn't let us. We foxed 'em, I betcha!"

One elderly couple I liked, attracted to them by their unassuming manner and their evident attachment for one another. The man could not read or write, but had earned a modest competence as a dairyman. After we became well acquainted, he told me his story. He was the eldest of a large family that lived far back in the mountains, where their father raised nearly everything they had. He gathered medicinal herbs for a cash crop, taking out two or three pack-horse loads a couple of times a year, to trade for salt, molasses, cloth, etc. On these trips he would never take any of the children. He was not sure why his father chose such an isolated existence, but had concluded he was wanted by the law for something. By the time this fellow was 16, he had never been more than a few miles away from the home cabin. On his sixteenth birthday, the father told the lad that he was a man now, and was on his own. Since he was on his own, he decided to go down on the flatlands, and see what it was like. He got a job on a dairy farm. The farmer had a daughter. They hit it off pretty well and wanted to get married. Her parents were willing, and it didn't occur to him to consult his parents, so it was arranged. The day before the wedding, her father took him aside and said, "Son, I know you, and I know how you were raised, and I know you have no idea how to treat a woman. But I've seen how you treat my cows, and if you treat my girl as well as you treat my cows, I'll never fault you."

"Well, I tried to remember this, and tried to live up to it, and by Jing it works. We been married nigh onto 60 years now and we are both satisfied."

For some reason I do not understand, color has a lot to do with the effectiveness of medication. After I noticed this odd fact, I made an interesting series of tests. Our contract called for us to furnish medicines, and of course we had many calls for aspirin and cough syrup. I bought aspirin tablets in four different colors, red, green, white and yellow. I also bought a colorless cough syrup, and used cake coloring to produce the same colors. When there was a call for either medication, I would give some of two or three colors, with a spiel about how I was trying out some different things and would appreciate it if they would report which kind worked best. People like to be helpful, and nearly all reported back. In almost every case, the red medicine worked best!

For a time, the Southern Pacific operated a quarry near Garibaldi, getting out rock for road ballast. They were working on a vertical face, about 125 feet high. Because the rock was badly broken up, pieces dropping out of the face were common, and a lookout was always stationed where he had a good view of the entire face and could call a warning whenever a piece started to fall.

One day, one of the quarry workers was brought to the office with the skin and muscle of one side of his right upper arm cut through above the elbow, and hanging as a loose flap from the shoulder. As he told the story, "The lookout gave an alarm, and I looked up to see a big rock coming right at me. I started to run, but decided I couldn't make it and threw my arm over my head. I just couldn't run quite fast enough."

On the way to the hospital, we passed the quarry and he pointed out the rock that had hit him. It was bigger than a boxcar. If he had been able to run fast enough to get three or four inches further away, it would have missed him; if he had been three or four inches closer, I would have had no job. If he had devoted all his energy to running instead of putting his arm over his head, he would probably have made it without getting hurt, but I guess expecting a man to make that decision in such a time of stress is asking too much of human nature.

Understandably, some of the funny situations arise about the use of laxatives. It seems to be a natural. One day a young fellow came into the office and wanted a laxative. He made it very clear he wanted something prompt, powerful, and effective. I fixed up a concoction and he downed it. Before he went out, I suggested he go right home. Next day he was back, and said, "Doc, do you know where I live?"

"Yep."

"You know how far it is from here to there?"

"Yep."

"Well, you missed it 20 feet." And he stalked out.

Tom Leslie was bald. Not just thin on top; not just a high forehead; not just a bare patch on his pate. He was *bald,* except for a very narrow fringe around the base of his skull, and three brave hairs standing upright on the top of his head. He was a clean man, but even the best of us has an accident sometimes. While Tom was in for an insurance examination, I noticed a louse climbing up one of those three hairs on his crown. It was fascinating. That louse slowly climbed up that hair, like an ant climbing a grass stalk.

By a gesture I had called the nurse's attention to the small drama going on over Tom's head, and she was watching as intently as I was. She could keep a hand over her mouth, while I had to carry on a conversation with a straight face.

Why was that louse out in the open instead of hiding in the fringe of brush just above his collar? And why was the louse on a clean man like Tom? Where and how did he pick it up? Did he have any more? Did he leave any of them on the chair in the waiting room, or would he leave any on the examining table? While pondering such questions and watching that louse, I listened to Tom's heart so long he became alarmed. He thought I must have discovered something terribly wrong.

I tried to take his blood pressure, but could not keep my eyes on the sphyg. That louse! Finally it reached the end of the hair and rested there, swinging back and forth like an acrobat on a swaying pole. The nurse was breathing hard but making no other sound. Her hand was still clamped over her mouth and she was staring so intently I was afraid she would attract Tom's attention. By a supreme effort of will, I was able to complete the examination, reflexes, abdomen, extremities, etc.

At last it was over; he put on his shirt and coat. He wore no hat. He thanked me for my services, said good day to the nurse and walked out, the louse still swinging in the breeze on that long, lonely white hair.

Ethical question: should I have told him? I don't know. Ask Ann Landers!

One man confided some of his domestic difficulties to me. He said, "My wife is always cleaning or straightening up. The kids and I have to take off our shoes and wash our hands when we come in the house. If the kids have been out playing, they have to take a bath and change clothes before they can come in the living room. I haven't dared smoke in the house for years. The kids can't have a dog or cat, because they would track in more dirt. Everything has to be in its place. If one of the kids comes running into the house to go to the toilet, and drops his coat on the floor, before he is out of the bathroom she picks it up and hangs it in the closet, looking like a martyr. Each coat must hang on a particular hook, and shoes must be neatly placed side by side in a certain place in the shoe closet. If one of the kids upsets a glass of milk or his soup, you would think it was a major disaster. That woman is so nice she is nasty!"

One night I was called to the office by the police. They had a chap who had been roughed up in a tavern brawl. With him was his buddy. Both were typical wino bums, middle-aged, ragged, dirty, whiskery, with stooped attitude and shuffling gait. The injured man had several lacerations which required sutures. As I was working on the job, I suddenly noticed the other man was helping, and doing it exceedingly well. He had washed his hands, was keeping the instruments in order, and anything I needed was handed to me before I asked for it. When the job was finished and the pair started to shuffle out, I asked the chap where he learned to assist in surgery. He straightened up, gave me a snappy salute that would have done credit to any outfit, and said, "Four years in the United States Navy Medical Corps, sir."

Then he slouched over and the two old bums shuffled out of the office.

Tom Bruce, a rough-cut individual, about 70 years old, who smoked a foul smelling pipe, and had smoked as long as anybody could remember, unexpectedly announced he was quitting. When asked why, he said, "Effeminate habit. Women can smoke nowadays, so I quit."

Among the people that came to Garibaldi was a young couple with a small boy, five or six years old. This little fellow was rather frail, but gradually improved. One day the mother came in to make her routine report. After the usual questions and answers she said, "You don't know how good it is to have a tubfull of dirty clothes to wash."

I must have expressed surprise at such an unusual statement, for she continued, "Never, before the last two weeks, has that little boy gone out to play and gotten dirty. You have no idea how good he looks when he is dirty."

I'm afraid many mothers do not appreciate some of their blessings!

Some of the occupational hazards of the area have been mentioned—logging, sawmill and farm accidents, but there were a few others worth mentioning.

Commercial fishermen had two rather common ailments. One was known as "fisherman's boils," an inflammation of the wrists, often going to running sores. Most fishermen wore heavy canvas raincoats or slickers, made of a heavy cloth with a waterproof coating. The sleeves of these would get frayed about the wrists from wear and these frayed edges would be frequently soaked with fish slime and salt water. The end result was a rough surface, almost metal hard, contaminated with filth. Continuous rubbing of this across the wrists was almost sure to cause trouble.

The other complaint was "fisherman's bottom," an itching, burning inflammation of the buttocks due to long sitting on the wooden seat of a boat, wet with salt water.

Daffodil pickers' sore fingers came to the office in the spring. There were a few people who raised daffodils for the market, mostly the big yellow King Alfreds. The blossoms were picked just before they opened up, packed in large cartons, and shipped to wholesale florists all over the western states. The pickers snapped off the stems with their fingers, and their hands were gummy with the juice. They developed infections about the nails, sometimes severe enough to disable for several days. Fern pickers had the same trouble.

By the middle thirties, a few people had started raising mink. The diet was largely fish, sometimes not very fresh. I was not sure whether the aroma that clung to workers around mink was from the animals themselves, or from the fish remnants they left about their pens. Mink cages were up off the ground, and some of the ranchers let ducks range under the pens. That helped a little bit.

Mink are vicious little beasties, and will bite. Attendants were supposed to wear heavy leather gloves, but sometimes they would forget. At times a mink would bite through the glove enough to cause a nasty laceration, and they were always contaminated. I saw one fellow that got too close to a mink; the animal jumped up and grabbed his nose, almost biting it off.

There were a lot of night calls in my type of practice. I developed a plan for leaving my clothes in a particular place each night, always arranged in the same way. Thus I was able to dress and get out with minimum disturbance of the family. I could get dressed without turning on a light or being really awake.

One stormy evening, I came home just about bedtime. On the way from the garage to the house, I slipped and set my backside squarely in a puddle of water. In the house the pants came off and were hung by the stove in the living room, while I went to bed. During the night, there was a call from a very outspoken neighbor. The pants were not in the usual place and I was so groggy with sleep I had forgotten what happened before. I rummaged around in the dark, trying to do it quietly, but no pants. After several minutes search, I remembered, retrieved the pants, and took off. The woman calling was indignant because it took me so long to get there. In an attempt to mollify her, I started to tell her the reason, but she let me talk just long enough to tell her I got my pants wet, when she broke in with, "Fine thing, big man like you wet his pants. When will you grow up?"

A tough young fellow who lived several miles out in the country had a severe infection of the leg. I saw him several times at his home and was making very little headway, if any. Hospitalization was advised, but he steadily refused. One day he surprised me by announcing, as I entered his room, "I want to go to the hospital."

"All right. Go in this afternoon."

It happened that the juiciest murder trial ever held in Tillamook County was due to start the next day, but I never connected the two events.

He checked in at the hospital, was most agreeable and cooperative. Next morning he demanded his clothes. Of course the nurse refused to give them to him without an order, and he raised quite an uproar about it. When I walked into the ward, he called me in a loud voice, and when I got near his bed he started to tell me what a stinker that nurse was. When he paused for breath, I asked what was the matter. "That old so and so wouldn't give me my clothes, and I don't know where she put them."

"You don't need clothes, you are going to stay right there in bed with hot packs on your leg."

"But I gotta get out of here."

"Why?"

"I want to go to that trial."

"You can't. You are staying here with hot packs."

And he signed a release, checked himself out and left.

Don't know whether he went to the trial or not, but presume he did. And the leg got well with no further attention from me.

More Tales

For a time, there were two butcher shops in Garibaldi. The proprietors were hard competitors, but the best of friends. One of them tells this story:

"One Thanksgiving, I ran out of turkeys. A woman came in and asked for a turkey. I said, "Just a minute," ran across the street to Bert's shop and asked if he had any turkeys to spare. He said he had just one left, and not a very good one, but I was welcome to it. I took that turkey across the street and showed it to the woman, but she didn't like the looks of it and bought a ham. Pretty soon here came Bert; wanted to know if I sold that turkey. He took it across the street, but couldn't persuade his customer to take it. In a little while I got it, but couldn't sell it. That went on all afternoon until finally one of us sold it. That turkey crossed the street six times before somebody took it home."

He also told me this: "A fellow in the neighborhood, Tubby Jones, raised some horseradish for a little extra money. He would clean it, grind it, and put it in mayonnaise jars—pure horseradish and pretty potent. I had a row of his jars in my shop. One day an advertising salesman came in. He was a city dude, the know-it-all type, trying to sell advertising in something I had never heard of before. While I was waiting on a customer, he nosed around the shop and noticed this horseradish. As soon as I was through, he started talking about it, how you couldn't get good horseradish any more, it was all diluted with ground up turnips and no real horseradish flavor. He wished he could get some real horseradish like his grandmother on the farm back in Ohio used to make, etc. I let him talk for a while, then said, 'This is pretty good horseradish.' He pooh-poohed the idea, said there wasn't any really good horseradish any more. I argued with him, and he wanted to bet me he could take a tablespoon full of that stuff at one time without anything else. So I rustled up a spoon; we opened a jar; he dug in and came up with a heaping tablespoon of horseradish. That stuff was condensed fire; Tubby wasn't smart enough to think of diluting it with anything. That salesman put the whole spoonful in his mouth, and in about two seconds a wild look came in his eyes; he rushed out the door and spit the horseradish in the street. I had to go out and help him back in the shop, for by that time his eyes were watering so he couldn't find the door. We went in the back of the shop; he bent over the sink, washed his mouth several times, then his face and eyes. When he could get enough breath to speak, the first thing he said was, 'That is real horseradish.'

"He paid his bet, then bought three jars to take home with him. I guess he liked horseradish."

Another of his tales: "For several years, Bert and I were in partnership. I ran the shop and he scouted around for meat. Buying from the big packers was expensive because of transportation. We could give the farmers a little more for their animals than they would get if they shipped to the big markets, and still get our meat cheaper. Bert would go to a farm, dicker for an animal, and if he could make a deal, kill it there and work it down to pieces that he could get in his truck.

"One day a farmer sent word he had a bull for sale. This farmer had the reputation of being a hard man to do business with. He was suspicious by nature, always looking for a crook. He was also exceedingly careful about small things and small amounts of money.

"Bert went to his place and took a look at the bull. He was a nice fat, big animal. After much haggling, a deal was made. Bert got his rifle from the truck and shot the bull as he stood in the corral. When he went down, he fell on the farmer's pitchfork and broke the handle. The farmer put in a claim for the value of the handle. Bert argued that the accident had happened when it was the farmer's bull, in the farmer's corral, killed with the farmer's knowledge and consent; that the bull fell on the pitchfork that had no business being there—it should have been in the barn. The farmer countered that since they had made a deal for the bull, he was Bert's bull; that he was responsible for everything his bull did, including falling down. In fact, he was directly responsible for the bull falling, for he shot him. Moreover, since he no longer owned the bull, and the animal was on his property, it was trespass. Furthermore, he would hold the carcass of the bull as security for the claim for the value of the pitchfork handle.

"Bert knew when he was out-talked and paid for the broken tool."

Pete Williams was an outspoken individual. He had a butcher shop at Rockaway, even then a well patronized summer resort. (Most tourists rented cabins and did their own cooking.) Pete carried a good variety of meat and fish, but his standby was hamburger. He made this himself every morning, sometimes several hundred pounds of it. One day a very fastidious, dignified elderly lady came in. She wanted hamburger. He motioned to a large mound of it in the case.

"Is it fresh?"

"Yes ma'm."

Again she looked at everything in the case and came back to the hamburger. "Are you sure it is fresh?"

"Yes ma'm."

And again she looked at everything and came back to the hamburger. Before she could ask again, Pete said, "Lady, if you want hamburger, you better buy it quick. It ain't gettin' a damn bit fresher."

In the same shop, a very proper woman who looked like a school teacher, came in and asked for *sal*mon. Pete said he didn't have any. She was disappointed that she couldn't get *sal*mon; she thought she could get *sal*mon any place on the beach. She had so looked forward to eating *sal*mon.

She came back the next day and asked for *sal*mon. The third day she was back with the same request. By this time the light was beginning to shine for Pete and he said, "Lady, are you by any chance trying to say *sammon?* If you are, there it is, I always have it."

Anyone who has ever cleaned a mess of clams knows that after they are cleaned and set for a short time, quite a bit of juice comes from them. It is called clam nectar, and is quite flavorful. Some dealers were not above adding a little water to it. Pete handled clams that were dug and cleaned on Nehalem Bay by a fellow named Chamberlin. One day a fussy man came in and wanted to buy some clams. Pete started to scoop up clams and nectar, but the customer objected. "I want clams, not that water." Pete asked what he was going to do with the clams.

"Chowder."

"You need this nectar to make a good chowder. Gives it flavor."

"Oh, I didn't know that. All right, put some in." And he took his little carton and went out.

Next day he came in and Pete asked him how he liked his chowder.

"It was fine. I want some more clams; we are going to have chowder again today. And be sure to put in plenty of that nectar."

Next day he was back again. This time he wanted half a gallon of the nectar. Pete asked why just the nectar.

"I dug some clams myself this morning and we are going to make chowder again. I want it to be good, so I want some of that nectar to put in it."

Pete solemnly sold him a half-gallon of the nectar.

A perfect memory I did not have and I envied those who did. One such was the manager of the sawmill. We sometimes played bridge with him and his wife. One evening a little argument came up. To prove his point, he took up the cards, dealt the hand as it had been, than ran through the play, trick by trick. Argument settled!

I told this incident to Roy Albers, the man in charge of the mill warehouse, who said, "I can match that story. A few days ago, I took a list of things we needed to Hans to get his O.K. on a purchase order. He picked out one item, a little bolt, less than an inch long, with a peculiar shaped head to fit a particular spot in a machine. 'Roy, you should have one of these,' he said.

'No, I have never seen a bolt like that. I have looked thoroughly, and there is no such bolt in stock.'

'We bought some of those some time ago, and we bought an extra to have on hand. You check your old purchase orders and you will find it.'

"I went back to my office and checked the orders. Sure enough, I found it. More than two years ago, before I was on the job, we bought two, and using the index number, I found one in the storage bins. I guess it is easy to see why he is boss, and I'm working for him."

One of the early sawmills was at Hobson Point, two or three miles down the bay from Bay City. The Company maintained a bunkhouse for the employees. At that time most of the transportation was by water but there was a trail that could be traveled by foot or by wagon in good weather.

One evening a bunch of the men were playing cards in the bunkhouse when one got tired of the game and decided to walk to Bay City to see if he could find some beer. There was enough moonlight that the trail was easy to follow where it was open but under the trees was rather dark. About half a mile along the trail dipped down into a deep gulley, through which flowed a little stream. Along the stream was a thick growth of alder, so the gulch was quite dark. As this man was carefully picking his way down the near slope he saw something white at one side of the road near where the trail crossed the creek. It seemed to be a woman, dressed in white, with arms upraised as if calling on the Heavenly Powers. Pretty soon she lowered her arms, knelt as if in prayer, then rose and extended her arms again. He thought of the tale, probably apocryphal, of a white woman who had been killed and beheaded by Indians many years before. This must be her ghost, calling to Heaven for vengeance. For several minutes he was too frightened to move, watching several sequences of calling on Heaven, then kneeling. At last his legs would obey his head, he turned and ran all the way back, breathlessly burst into the card game with the statement he had seen a ghost. As soon as he got his breath he elaborated at length on what he had seen. Some of the boys were skeptical and wanted to investigate. Others refused to leave the bunkhouse. Finally a small group of the unbelievers took lanterns and set out. When they got to the gully they found an ox that had gotten out of someone's corral, quietly eating the grass along the creek. This was a very big, black white-faced animal with long white horns. When it raised its head its poll was about five feet from the ground, and the horns were like upraised arms. Probably in the dim light it looked like a supplicating woman. When it lowered its head to graze it vaguely resembled a woman kneeling.

Of course, the man who saw the ghost came in for a lot of kidding by fellow workers. They named the gulch Ghost Gulch, and the name stuck. The little creek became Ghost Creek, and a deep hole where the creek flowed into the bay became the Ghost Hole. As such it is still known to fishermen as one of the best fishing holes of the lower bay.

Ed Pawley was superintendent of a construction outfit. He had come up the hard way; had very little education, but a lot of drive, ability and practical experience. When I knew him, he was probably near 60 years old, a burly, red-faced, cauliflower-eared flat-nosed Irishman, with a voice like the bellow of a bull. He was a typical tough, hard driving construction stiff who had managed to fight his way to the top of the heap.

My older boy, then in high school, got a summer job with the outfit. One evening he came home goggle-eyed. He said that afternoon, just before quitting time, Ed came around to check the job and started to yell and wave his arms as soon as he got out of his car. "No, no, no, no! That isn't right! This floor has to carry heavy machinery that must set exactly level. Those sills are not level and they won't do."

The foreman said he had just checked and they were level. Profane old Ed told him off in no uncertain terms, ordered all work stopped and the instruments set up to check the level. And it was off, just a quarter of an inch in 40 feet!

Hulda Gusty was an independent soul, fortyish, a large, strong, active woman. Both she and her husband were enthusiastic hunters. They always went deer hunting each fall, and both usually brought home the limit of meat, which helped to feed their flock of kids. One fall, Hulda got an elk tag in the annual drawing, but Gus didn't get one. She tried her best to induce him to take a few days off from work to go with her on the elk hunt, even if he couldn't shoot one. This he refused to do. So, after quite an argument, she threw the camping gear in the pickup and took off by herself.

It was late in the fall and the hunting area was rather high. When she got to it, there was over a foot of snow on the ground. About 50 hunters were on hand, but she was the only woman. She made her camp and took her part in the organized drive across the mountain. A few days later she was telling a group in the store all about it. A little old-maidish clerk said, with a simper, "I'd think you would have been afraid to be away out in the woods alone with all those men."

Hulda looked at her a minute, then said, "Gal, it was damned cold up there."

Incidentally, she brought home an elk.

Cougars

The best farm land in Tillamook County was used for dairying, but there were a few hardy souls, known as stump ranchers, who were trying to make a living among the stumps of recently cut-over land. They lived mostly on potatoes, pumpkins or beans which they raised between the stumps, with an illegal deer carcass at times, and fish out of the nearest creek. For cash they raised a few beeves or sheep, cut firewood, split shakes, gathered fern and foxglove, peeled chittim, or held down a job somewhere. Some who lived in the low land were known as "Swamp Angels." It was a precarious existence, but some men liked it. Don't think most of their wives shared that feeling.

Tom Mitchell was such a fellow, a tall, lanky guy, about 35 years old. One day he came in to talk about his baby and stayed to visit a bit. Ultimately he told this tale.

"My milk cow went dry about three weeks ago, and I arranged to get a couple of quarts of milk every day for the kids from Hans Schmidt, who lives about half a mile from me. The path to his place goes through the woods most of the way. One place it goes under an old windfall that has lodged about six feet above the ground. Yesterday evening I went over to Schmidt's to get the milk and was coming back when it was getting dark. As I got to that old windfall, I looked up just in time to see a big cougar stick his head over it. I don't know which of us was most surprised, but I moved first. I hit him square in the face with that bucket of milk, then took my foot in hand and left that place. Don't know whether he chased me or not—I couldn't have heard a team of horses. If he did, he couldn't run fast enough to catch me. My kids didn't have any milk for breakfast this morning, for I sure wasn't going to travel that path again in the dark."

I don't think there were many cougars in the country, but they were much feared. They took toll of calves, sheep and pigs, but I never heard of one making an unprovoked attack on a person. The so-called attacks were actions to get food, or to get out of what seemed to the animal to be a dangerous situation. The tale told by the wife of a stump rancher illustrates the first.

"Tom, my husband, had left home for Cloverdale, where he had heard there was a man buying chittim bark. We had a lot of it around our place, and Tom figured on peeling some of it if he could get a market. It was a good half day trip and Tom left about noon. He figured to get pretty close to Cloverdale, then camp, and catch this buyer first thing in the morning before he could get away from home. He took the rifle, for he thought he might see a deer on his way home.

"He had killed a deer a few days before, and there was still one hind quarter hanging up out in the shed. It was a hot afternoon, and I got to thinking about that venison. As hot as it was, that meat might spoil, and that was all the meat me and the kids had until Tom could shoot something else, or catch some fish. So, I got that hind quarter off the hook, cut the meat off the bones, and put it in a big pot to cook. Also, I cracked up the bones, put them in the soup pot, and set it to simmer. I guess the smell of all that cooking meat must have reached quite a ways.

"Come evening, I had the kids in, fed and bedded down, when I heard a cougar scream out in the woods. That is a sound that fair sets your teeth on edge. I closed and fastened the shutters and the door. Didn't like to do it, for I had to keep the fire going, and it was hot in that cabin, but I did it. Since I had the fire going I figured I might as well heat some water and take a bath when I was sure all the kids were asleep.

"Pretty soon I heard something moving around outside. Around and around the cabin it went. I thought of our old sow out in the pen, with her litter of little pigs, but there was nothing I could do about it. I was looking after my kids; she would have to look after hers. She knew there was something wrong, for I could hear her making a fuss, and I guess the cougar heard her too, for next day we didn't find any tracks near the pen. He evidently didn't want any truck with that sow, for she was a big one and could put up quite a fight.

"But the cougar kept going around the house. Finally he started sniffing around the door. Guess he had decided that good cooking smell was coming from there. Then he began to claw at the door, and it wasn't long before he could get his claws in the cracks between the boards. Up to that time, I was annoyed and mad because I was afraid the disturbance would wake the kids, but when I saw those claws come through the crack, I was scared. Tom had the gun; all the weapon I had was the butcher knife; and I sure didn't want to get close enough to that cat to use a knife. When he got his claws in the crack, I knew it wouldn't be long before he could rip a plank off the door. I didn't think he could get through the space of one plank, but if he could get one off, he could get two, and then I knew he could get in. Pretty soon he had that first plank off, and stuck his nose through the hole. It was almost big enough for him to get his head through, but not quite. While he was trying to get through, I thought of that kettle of bath water, boiling by this time. I picked it up and emptied it on that nose sticking through our door. He yowled, backed away, and I heard him scuffling around outside, but I was too scared to open the door and look out. Pretty soon all was still; I didn't hear him leave, but the old sow got quiet, so I was sure he was gone.

"Then I discovered I had picked up that pot of boiling water with my bare hands and had burns on both palms. But that was a cheap price to pay to get out of that mess. And when my hands hurt, I'd wonder about that cougar's nose. Maybe I got his eyes, too, because I could see both of them when I threw that water. And he was just looking for something to eat. Maybe he had some kids that were hungry."

Another stump rancher told me this:

"I live pretty far back in the hills, but I have a close neighbor. Our houses are on opposite sides of a deep canyon. This canyon is grown up thick with alder trees, some of them pretty good size. To make it easier for us to get back and forth, we cut a trail through those alders, dug out the sides of the canyon a little, so it would not be too steep. It was nearly impossible to get off the trail, for the underbrush on each side was so thick, and all around were those big alder trees, their branches meeting overhead.

"One night I had been over at Bill's house. We had played pinochle until nearly midnight when I started home. There was a full moon and no clouds. Even under those trees it was not really dark.

"About the time I got to the bottom of the canyon, I heard a sound like a cat spitting, looked up, and, on a branch of an alder, between me and the moon, was a cougar. He looked as big as a cow to me. My hair stood on end till it pushed my cap off my head. I had always heard that if you face a wild animal he will not attack. Could not think of anything else to do, so tried it. I turned towards him and backed slowly away in the direction of home. Mr. Cougar gave a big yowl, dropped out of the tree and out of sight, but a minute later I spotted him in another tree, closer. I kept on looking at him and backing away. Pretty soon he repeated the maneuver, and I kept on looking at him and backing away. By the time he had climbed into four trees, I had reached the top of the canyon, and my house was only a couple of hundred yards away. I had a dog, a big, friendly mutt of no particular use except that he had a loud voice to scare the deer away from my garden. I thought if I could get him to create a bit of diversion, I could get to the house pretty fast, so I whistled for him. He came lolluping up, apparently not realizing there was anything wrong. As he got near me, that cougar gave a scream; the dog looked up, saw that cat, tucked his tail between his legs, and ran faster than I ever saw him move before. The cougar jumped for the dog, but by the time he hit the ground, the dog was halfway to the house. He ran after him, but that dog was too fast, so he went on into the bushes. The dog ducked under the house and he never came out for three days. It was weeks before he would go near that trail across the canyon.

"Another neighbor had a pack of cougar dogs. I told him about my experience next day, and he came over. His dogs picked up the trail and treed the animal after a mile or mile and a half chase. After it was dead, it didn't look nearly so big as it had the night before. I wonder if it was the same cat."

A Tillamook family liked to go camping. They had a crude cabin away up the Trask River, considerably above where the park now is. To get to it, they could use a team and wagon for a few miles, then they would make packs for the horses and themselves and go up a rough trail to the cabin. They stayed there two to four weeks every summer.

One time when they got to the cabin, it was late afternoon of a rather warm day. They were hot and tired from the long walk under packs. While the man gathered sticks to build a fire, the woman took a couple of buckets and started to the spring, followed by the boy, about 12, with a couple more buckets. They were going to heat some water and enjoy a good wash-up.

Brush had grown up along the trail to the stream since they had been there a year before, in some places almost closing it off. Suddenly there was a scream and something hit the woman on the back as she was bending over to dip up a bucket of water. She turned around, grabbed the boy, and started to give him a good, workmanlike spanking, saying, "I'll teach you to scare me that

Typical Tillamook County terrain of period. Note plank road. (OHS)

way, to yell and hit me on the back with a stick. You may think that is funny, but I'll take the fun out of it."

Then the man came running up with his gun in his hand, much puzzled as to why she was whipping the boy, and asking, "Are you hurt?"

When the woman cooled off a bit, the man said, "I saw a cougar jump across the trail and I thought he jumped on your back. Let me look." And sure enough, her dress was torn and there were scratches on her back.

When they looked about, they found where a cougar had been lying on a lodged tree, not more than two or three feet from the ground where apparently the cougar had lain a great many times, for there were many scratch marks and hairs sticking to the wood. The woman had evidently surprised the animal and in his efforts to get away he had made these scratches on her back.

The boy got an apology. One of the first chores was to slash out that trail. They heard a cougar scream several times that summer, and once they heard a deer being killed, back in the timber, but they did not sight a cougar again that trip.

People and Changes

Probably as significant as any change of the last half century, is the change in bathing habits. When we went to Tillamook, only a very few homes had bathtubs, though a great many had running water. In a country as rough and well watered, it was usually no problem to have water in your house—you just went a little way up the nearest canyon, threw up a small dam, strung out a line of pipe, and you had running water, with plenty of pressure. Sometimes two or three families would put in a partnership job. Tillamook city had a municipal supply, possibly others. Most of the communities smaller were served by privately owned systems.

The dam made a little reservoir that served as a settling basin to allow suspended matter to settle out; some had a second basin to do a better job of removing sediment. Chlorination was begining to be adopted elsewhere, but as far as I know, was not used in our area until the late twenties. Watershed protection was not a factor; these little reservoirs were usually in the woods where people seldom went and the only animals were the wild life.

If a logger started to work in the valley above your system, it really raised hob with it. The severe disturbance of ground cover and tearing up of the surface soil caused extensive erosion, with muddy water below. The settling basin soon silted up, and the finer particles of clay settled out very slowly.

In spite of the fact that most people had running water in the house, tubs were scarce. Many people had hot water on tap. They had the tank alongside a wood stove, with coils in the firebox. To have hot water, you had to have a fire in the stove. That was easy enough in the winter, when you had a fire in the stove for warmth anyhow, but if you wanted hot water in the summer, you had to build a fire. Some of these systems were connected to the cookstove, so the fire used to cook a meal heated water to wash dishes. But the coils took up considerable room in the firebox, so the cookstove wood had to be cut smaller and the fire stoked more frequently.

But to get back to tubs, I do not know why they were so slow coming in. People could bathe in the creeks in the summer time; usually there was one handy, and in the winter the family washtub was brought in, filled with water from the tank if there was one; if not, water heated on top of the stove. Frequently the same water served several members of the family.

For some people, all this was too much bother, and before the cool weather was far advanced, your nose enabled you to identify them. Lack of interest in bodily cleanliness usually was coincidental with lack of interest in clean clothing, and that soon led to habitation. One fellow told me that the most plentiful and hardest hunted game in his area was greybacks. Certainly they reproduced rapidly where conditions were favorable. There was no closed season on them, either.

At this same time, baths were a profitable side line for barbers, and nearly every shop had them. As home tubs increased, the barber shop tubs got less use. I haven't seen the sign "BATHS" in a barber shop for many years.

Showers were common in logging camps, but not in homes. I do not know when the camps first put in a bathhouse with showers; they all had them by the time I knew about them.

Indoor toilets came in before tubs. There were many old-fashioned privies in 1923, but they rapidly went out of the picture.

Sewage disposal depended on location. In some cases, sewer lines emptied directly into a stream. Farmers that had liquid tanks sometimes connected their house sewer line to them. Cesspools, and septic tanks with field drains, were the usual methods of sewage disposal. In areas where the water table is near the surface, that presents problems. Sewage collection systems for the larger communities finally were built, but sewage disposal plants were very slow in coming.

Sewage from Tillamook city emptied into the Trask River, and there were always a lot of fine fat fish around the outlet. For some reason, local fishermen did not seem to care to fish there, but when outside fishermen were directed to that place, they usually did well. Nobody bothered to tell them about the sewer line, and if they did not happen to notice it, they were quite pleased with Tillamook fishing. Anyhow, everybody cooks fish before eating them.

A treatment plant has been built in late years. I wonder if it spoiled that good place for tourists to catch fish?

The flu epidemic hit Tillamook just as it did the rest of the country, and the doctors had just as wide a variety of treatment as they did elsewhere. One of them had the notion that bathing was a bad thing for a flu patient, and strictly forbade it, no matter how odoriferous the patient became.

He was treating a little boy whose mother was one of those excessively clean women. Her house, her clothes, her family, were washed frequently and vigorously, whether they needed it or not. Naturally, she did not like the "no bathing" edict for her boy, and every time the doctor called, she needled him about bathing the lad. One day she asked the usual question and got the usual answer. She said, "But Doctor, he *stinks!*"

That was too much. The old fellow had had more than he could take that day and exploded. "Hell's fire, woman, he'd better stink alive than stink dead. Let him alone." And he stomped out.

During the days of the Civilian Conservation Corps, a camp was located a few miles out of town, employed in clearing the right-of-way for a new road across the mountains. The first lot of boys there were from the slums of Baltimore. There were some good boys in the lot, as well as some of the other kind, but at first they were a confused, homesick bunch of kids, suddenly transplanted from the crowded streets of a big city into the big timber, doing strange work, under conditions many had not even seen in moving pictures.

A local man, who had been a major in the army, discovered that the superintendent of this camp had been his former top sergeant. While visiting with him one day, he asked a question about discipline.

"Hell, Major," was the reply. "We ain't got no discipline. We use diplomacy."

From what I hear, I don't think the Job Corps has improved much on that formula.

One boy was telling, with great glee, about the "big woim" he had captured. When he pulled it out of his pocket, it proved to be a two-foot snake. Fortunately for him, there are no poisonous snakes west of the Coast Range!

A few of the boys married local girls, settled down, and became good citizens. Some openly held local people and affairs in contempt—they were "hicks." A few got in trouble with the law. Most of them quietly served out the time of their enlistment, worked more or less faithfully, ate regularly and heartily, participated in camp activities, and had little to do with local people. Several attempts were made by local people to get acquainted, but they never seemed to get anywhere. No common ground, I guess.

I have often wondered if the year in Oregon had any real effect on them, or if they took up their old life where they left it. Now we see the same old idea warmed over in the Job Corps. I wonder if the present organization has derived any advantage from the experience of the previous one. It would be almost too much to expect.

Josh Goodwin was older than most of the people that moved to Garibaldi. He claimed he couldn't work, and maybe he couldn't. At least he didn't hold a job. He limped badly when he walked, and leaned heavily on a cane. He had some boys at home that worked in the mill, and his wife was always available to help when there was sickness in the home, or to do baby sitting (we didn't call it that in those days). Maybe he just didn't have to work. One thing he did do was tend the family garden. He evidently liked to garden, and he always had a large, well cared for, beautiful garden.

One night a neighbor's cow got loose and broke into his garden. One couldn't blame the cow, that garden must have looked like a cow's idea of heaven, with its beautiful corn, peas, cabbage, lettuce, etc. At least, the cow made the most of her opportunity and by the time Josh discovered her next morning, she had devastated that garden.

Josh was furious, and had reason to be. He was so angry he forgot he needed a cane to lean on. He chased that cow down the street at a dead run, abusing her verbally at the top of his voice, and whenever he could get close enough, whacking her rear end with his cane.

After that, we sort of questioned his alleged disabilities!

Al Swett, an old time Tillamooker, remembers this one.

"After we got a golf course at Alderbrook, near Bay City, I got to playing and thought I was pretty good. I had a good drive; I could really wallop it, but when I got near the green, I wasn't so good; in fact, I was lousy. One day at Lions' Club, one of the boys brought as his guest a salesman for fishing tackle. We got to talking about sports and I did a little bragging about my golf game. The visitor did not seem much impressed. The upshot of it was, we cooked up a match. I was to shoot a golf ball in the usual way; he was to use a casting rod, and cast his way around the course, each cast to count as a stroke. We made a sizable little side bet.

"I felt fine. I thought I just as good as had his hide already nailed to my barn door. I knew I could drive several times as far as he could cast. What I didn't know was how accurate he was. When that guy got 60 or 70 feet from the cup, he was just one cast from being in. It was uncanny. I'd get in the neighborhood of the green in maybe two strokes, and he would be near me with four or five or six. But I'd monkey-doodle around on the green, and waste my advantage; then he would give a flip with that little stick, and Wham! his plug was in the cup. It got my goat. We ended up exactly even. I still don't quite see how he did it, but I know he turned in an honest count for I had one of my friends checking up on him."

Matrimonial agencies have been the subject of considerable levity, but maybe there is a good word to say for them. I have known two men who got wives that way, and both marriages turned out well. One fellow was a Swede logger, probably in his late thirties, not very prepossessing to look at, but honest, hard-working and kindly. He got a wife somewhat younger, not very attractive, but possessing many basic virtues. They had a large family and seemed to be a happy, close-knit, cooperative unit. As far as I know, none of the kids turned out to be a genius, but they all developed into good citizens.

The other fellow was a Swiss dairyman, a bachelor in his late forties. He had very little education, had lived alone since his mother died more than ten years before. He had milked cows all his life. He married an old maid, city school teacher, of about the same age, a woman of considerable intelligence and refinement. Everybody was curious as to how she would take it, living on an isolated farm, adjusting to a confirmed bachelor, far below her in intelligence and education, and living in a house that had not seen a woman for many years. But they got along fine, seemed very happy together.

These women were known in the community as "Montgomery Ward wives." It would seem that that company need not be ashamed of those sales.

In the case of the second couple, to everyone's surprise (and theirs, too, I think), she soon became pregnant. When I was able to be sure about it, she told her husband, and at once he came into the office to get the information confirmed. Maybe he thought she might be spoofing him. When I backed up her story, he said, "You know, my wife, she funny. She just as glad as I am."

She went through the pregnancy without trouble, but had a hard labor. As she was coming out of the anaesthetic, she shouted over and over. "Who says an old lady can't have a baby? Who says an old lady can't have a baby? Who says an old lady can't have a baby?"

There was a Swiss man in the community, a prosperous farmer, a very large man, a friendly, jolly sort of chap with a big booming voice that could be heard a quarter of a mile away. He was frequently in the office, and was always a welcome visitor, for he was intelligent and interesting to talk to, even if sometimes he was a bit difficult to understand when he talked fast.

One day, just before Christmas, as office work was slack, my assistant decided to do some Christmas shopping. She changed from her white uniform to street clothes and went out. In a few minutes she was back, her face a flaming red. To my question as to why she was back so soon, she said, "Just as I got down by the bank, where the street was crowded with shoppers, I met Mr. Sucro. I said, 'How do you do, Mr. Sucro,' but he didn't answer—did not recognize me, I guess. But as I walked by, he kept turning around and looking at me. When I was about half a block away he called out, in that big voice of his, 'Now I know who you are. I never saw you with your clothes on before.' At least a hundred people who know me heard it, and I just couldn't face it. So, I came back."

Bar View is a small beach town at the mouth of Tillamook Bay. At one time there was a dance hall there, and they had a dance every Saturday night during the summer. One year a dance orchestra of five members, from Tillamook, was engaged to furnish the music. They arranged with Martha, a girl in Tillamook, to furnish them transportation. This lady was a big, fat sloppy gal.

U.S. Coast Guard Life Saving crew drilling above, Cap Farley holding the steering oar. Below is view of CG station and Bar View, circa 1930s. (Wesley Andrews Collection, OHS)

Her boyfriend had let her use his car while he was working at a camp pretty far back in the timber. It was a big car and she used it for a sort of unofficial taxi service. Each Saturday evening, she would take the orchestra to Bar View, hang around the hall, sometimes picking up a local fare, then take them home after the dance. That was when the roads were bad, and getting that big car stuck in a mudhole was a common event. When it happened, everybody got out and pushed and lifted until the car was free. Martha did her part of this, maybe more, for she was stronger than most men.

One night, near the end of the season, they were stuck unusually tight. Everybody was out, doing his best, when suddenly Martha gave a big grunt and said, "Ooops, here it comes," went over to the grass by the side of the road, and delivered a fine, big, boy baby. None of her passengers, who had been riding with her all summer, had ever suspected that she was pregnant.

By the time the boys got the car out of the mudhole, she was on her feet, had wrapped the baby in a spare coat which she had with her, got in the car and drove home. The next week she was ready to make the trip again.

A Tillamook car wrecker told me the following:

"One morning I got a call from a rooming house across town. The woman calling asked me if I wanted to buy a hearse cheap. I didn't have many calls for a hearse, but thought if it was cheap enough, I might be able to work it over into a pickup and make a few dollars. When I got there, it was standing out in front, and it was a beautiful vehicle, one of the best makes; it seemed to be well cared for, not many miles on it, good tires, and more fancy work than I have ever seen on one rig before or since. But something inside me said, 'Be careful.' I went up to the door and found two dames in the parlor. One gal, who did the talking, was a large, jolly, laughing individual, who, apparently, never saw a stranger. She offered the car for a ridiculously low price, less than I could get for the tires on it. But that something still said, 'Be careful.' After I talked to her a while, I asked her where she got that hearse. She gave a big laugh and told this story, stopping every little while to slap her leg and laugh.

'Me and Sally here, we live in Astoria, doin' a little of this and that in one of the houses there. We make pretty good and had a little stake saved up, so decided to go to Portland and have ourselves a time. We went and we had it. After about three days, we were down to two or three bucks and no ticket back to Astoria. Then I thought of a friend of mine that had a used car lot, so I called him up. This was some time after noon. He was glad to hear from me; said come on over to his apartment. Business was pretty slow and he would take the rest of the day off, and we'd celebrate. He had picked up a friend, and the four of us fooled around his apartment, doing this and that, drinking a little hooch, playing pinochle, just having a good time. Come evening, Bill went out and got some steaks and I cooked 'em, fried some onions, etc., and we had a good dinner. Of course we still had a little redeye left and we tapped it once in a while, but we weren't drunk, you understand, just feelin' good and havin' a good time. Sometime after eleven o'clock, I pretended to notice the time, jumped up, and said, 'My goodness, I've been havin' such a time I forgot to notice the clock. We have to get back to Astoria and the train has gone. Sally, what will we do?'

'Bill said, "Oh, never mind, we can take care of that. I got a lot full of cars downstairs; you can just take one of 'em and drive home. You can drive, can't you?" I said sure, so Bill said, "All right, let's go down and pick out one."

'We went down to the lot, and sure enough he did have a lot of cars, a hundred or more, I think. Bill said, "There they are, take your pick." So I started lookin' around and spotted that hearse. Ain't she a beauty? Bill said, "All right, she is yours." So me and Sally got in and struck out. Somehow we got confused, with lights and all, and got out of Portland on the wrong road. When we discovered we were on the Tillamook road, I said to Sally, 'That doesn't matter, we will go to Tillamook; my aunt Maude lives there. We'll stay all night with her and go up the coast in the morning.' It was about midnight, or a little after when we got to Newberg, and just out of Newberg we caught up with a bindle stiff who looked pretty tired, so I asked if he wanted a ride. He said he sure did; he was headed for Tillamook and that was a long walk. I told him there wasn't any room on the seat, but he could get inside, which he did. He unrolled his bindle, pulled a soogan over himself, and went to sleep.

'Sally was snoring in the seat beside me, and I was gettin' pretty sleepy myself. Maybe that's the reason I forgot to watch the gas.

Anyhow, we ran out of gas right smack in the middle of Sheridan, and I coasted up to a gas station, but it was closed. In fact, everything in Sheridan was closed. While we were sitting there wondering what to do, the night watchman came up and asked us what was the trouble. I told him we were out of gas, and we just had to get to Tillamook. He was a nice guy, and offered to wake up the fellow that ran the station. After a while the guy came, still half asleep. I thanked him for coming, but he said he was always glad to help keep an emergency vehicle running. Civic duty, and that sort of stuff. He was a nice guy, too. I had figured out how much gas my money would buy, and told him to put that much in the tank. Just as he was putting the gas tank cap back on, that bindle stiff inside the hearse waked up, sat up and looked out to see what was going on. The gas station man saw him, turned white, and took off. I though it was real nice of him not to charge an emergency vehicle for gas, and drove on.

'When we got here it was daylight, and the men were going to work. When they saw a hearse pull up, several came over to see what had happened. While they were standing around, that bindle stiff waked up again, and you should have seen them scatter! Aunt Maude gave us a bed and we just waked up.

'So that's how we got here, and that is why I have a hearse to sell. That old buggy handles like a truck, and I wouldn't try to take it over that beastly road, with all that puncheon and curves, between here and Astoria, for anything. All I want out of it is to get to Astoria, and if you will buy two stage tickets to Astoria, you can have it.'

"I didn't buy it; I was afraid that guy Bill might feel differently about the transaction, after he had a good sleep and his head cleared a bit. Sure enough, that is what happened.

"Don't know how those gals got to Astoria."

There were a few women in the community that were available to help in a home if the mother was out of circulation. One of these was Ma Taylor. She was a big, strong woman with a rough manner, a sharp tongue, and a heart as big as a pumpkin. If anyone needed help, she was there, with never a question about pay.

I was making a call-back on a baby case the day after delivery, and just as I came through the front gate, Ma Taylor was escorting the father through the front door, one hand on his collar, the other on his belt. When she got him to the edge of the porch, she kicked him in the place nature provided for that purpose. He landed in a pile at the bottom without touching one of the five steps on the way down. As she dusted off her hands, she said to him, "When you sober up, you can come back in and not before."

That woman had an uncanny ability to judge the time of a delivery. She must have cared for more than a hundred cases for me; I was never too late and never had to wait more than 30 minutes.

There was a large shiftless family in the neighborhood. The youngest, a lad of six or eight, came down with pneumonia because he did not have warm enough clothing for the season. Nobody was giving him any care, so we passed the hat and hired Ma Taylor to care for him. She went to the home, and was doing the best she could for the boy, but the older members of the family bothered her so much with objections and suggestions that she finally got fed up. She found a piece of 2 x 4 about three feet long, and announced to the assembled family, "If any of you blankety blank lazy bums want to stay in here and help me take care of Johnny, fine, but the first soandso that comes in this house without bringing in something useful, or doing something useful, gets his head broken. Now out, every stinkin' one of you."

They nearly tore the door off the hinges getting out. That club stood in the corner every time I called to see Johnny, and so far as I know, not one of that tribe poked his nose in the house as long as she was there. I have no idea where they slept; it was too wet to sleep in the woods. But that was their problem, not mine.

At that time and place, one of the usual treatments for acute respiratory infections was onion poultices. Of course, it was not as effective as the modern anti-biotics, and now is seldom used. In fact, I think the art of making and using an onion poultice is rapidly dying out. Just on the chance that therapy again comes a full circle, and the lowly onion poultice be found to have virtue, I think it might be well to set down directions for making and using it. I wouldn't want posterity to be deprived of the accumulated experience of our time because of my indolence or thoughtlessness.

Make four flat muslin bags, large enough to cover the patient's chest. Fill the bags with sliced onions and sew them shut. Some insisted that the onions be fried before being placed in the bags, but that is not necessary. Place the bags in an oven and cook until the onions are soft and mushy. Then put one bag, while hot, on the front of the patient's chest, and another on the back. Fasten them in place with a binder, preferably heavy flannel. As soon as the poultices cool off, replace them with hot ones from the oven, and put the cool ones back in the oven.

A grandmother skilled in the use of onion poultices was a wonderful asset in handling a pneumonia case. Of course, the onion aroma permeated the entire house, and could be detected weeks later. That was expected. If the passerby on the road in front of the house could smell the onions, granny was doing a good job!

Once I was called to see a man who had been thrown by a horse. He had several broken ribs. He was a large, muscular man and his wife was a large, fat woman. I'm no featherweight. This man was lying in the middle of a double bed. I sat down on the edge of the bed; his wife sat on the opposite. To strap him, I had to lean far over. The wife leaned over from the other side to help. This was too much for the bed; the rails spread, allowing the slats to slip off, some on one side, some on the other. The net result was that the mattress formed a "V" groove, with the injured man in the bottom, and the wife and me both lying on top of him. I do not think it is a good thing to do to a man with broken ribs—at least he didn't like it, and told us so very clearly and emphatically. He never called me again, either, so I guess he meant what he said.

One chap was using a barn broom to sweep out his truck. When he finished, he threw the broom to the ground and jumped down. Just how it happened, he didn't know, but somehow the broom bounced so that he came down on the end of the handle. This handle tore its way through his trousers, through the left scrotum, and up the inguinal canal. Judging by the blood on the handle, it had penetrated almost twelve inches into his abdomen. The broom had been used around the barn for several months and was grossly contaminated by manure and all the other kinds of dirt found around a barn. Because of the severe contamination, I was afraid to close the wound, but it was necessary to do something to keep his intestines from coming out. He had held them in with his hand all the way to the hospital. Sulfanilamide was just coming into general use, but we had not had enough experience with it to know just how it would perform. Since I could not think of anything better to do, I packed his wound with sulfanilamide, wound on a tight bandage and put him to bed. The only complaint the young fellow ever made was about staying in bed, and that the bandage was too tight, but I insisted that he stay put and that the bandage be not disturbed. The truth is, I was afraid to take it off. After two weeks, he insisted on going home, so with fear and trembling the dressing came off. I had visions of the lad's guts rolling out in bed behind the bandage, but nothing happened. On examining the wound, it was evident the sulfanilamide powder had hardened into a solid concretion, plugging the hole like a cork plugs a bottle. After trying various things to get it out, I finally had to take a screw driver to break it up. There was solid healing behind the concretion. The boy went home without a hernia, and feeling fine. There were no ill after effects.

An interesting sequel came several months later while I was in Chicago. One day I attended a surgical clinic in one of the large teaching hospitals. Between cases, I reported this incident. The surgeon was polite but evidently skeptical. His manner said quite plainly that he thought nobody but a hick would even think such a story might be believed, and some of the students present tittered.

The next case was listed as a recurrent cancer of the rectum. The patient had been operated on originally in a famous eastern hospital, and the record had been obtained by the Chicago hospital. It went into detail about the technique of removal and packing the wound with sulfanilamide. When the patient was brought in, tumor masses about the anus were plainly visible. The first step was to remove a piece of tissue for quick microscopic examination. When the surgeon cut into one of the tumors, there was a grating sound and a grayish mass was exposed. He laid down his instruments, got up from his stool, walked over in front of me and said, "Doctor, I apologize."

Mr. A. S. Bent had a little store in Garibaldi. He was just across from the mill entrance, and sold a few groceries, tobacco, candy, work clothes—such things as the men would be likely to pick up as they went by. He was an old man, probably well along in his seventies, short, rather fat, with billy goat chin whiskers. He was a Tennesseean who talked with an accent that set him apart from the rest of us. In addition, he told many tall tales that were

Top, boating at Tillamook. Center shows Bay Ocean Natatorium, c. 1920s, later claimed by ocean along with many houses. Below, not for timid. (OHS)

Foot bridge at
Mouth of Salmon-berry

pretty hard to accept. The net result was that we thought he was a big windbag. What made the situation particularly hilarious was that Daddy Bent was a deputy sheriff.

Nate Needham was a logger, a big, red-headed, powerful man, rather erratic in his talk and actions. One day he went violently insane and his tantrums terrified the community. Women and children stayed inside and kept the doors locked. The men gathered in groups to go to work. A warrant was out for his arrest, but he was so cunning that the officers were not able to catch up with him for a couple of days. Then someone spotted him sitting in a car in the mill parking lot, and called the sheriff in Tillamook. The sheriff, who was a friend of mine, called me, saying Daddy Bent's phone was out of order, and would I please tell him Needham was reported in a car in the mill parking lot, near his store. He (the sheriff) would be there as soon as possible, but in the meantime he should detain Needham in any way he could. As I was walking to his store, a few hundred yards from my office, I was chuckling about what a fix for the old windbag to be in, and wondering what kind of an excuse he would come up with to get out of the assignment.

Mr. Bent had not heard of the situation, for some unknown reason, and didn't know Needham, so I filled him in, and the tale lost nothing in the telling. I made Needham just about the biggest guy since Paul Bunyan, and the meanest since Captain Kidd. Then came the punch line: "Sheriff Aschim just called; said your phone is out of order and for me to tell you to detain this man until he gets here."

The old man never hesitated a minute. "Where is he?"

"Sitting in that car in the parking lot," which we could see through the window.

"I'll go git him," was his immediate reply, and he called through the back door to his wife to mind the store, threw his apron over one shoulder, revealing that he had neither belt nor shoulder holster, marched across the street, hauled the surprised Needham out of the car, and stood there holding him by one arm until the sheriff and two carloads of deputies came roaring up.

Were my feathers drooping!

Years later, after Mr. Bent was dead, I was talking to Jack Aschim and brought up the foregoing incident, telling him how surprised I was at what Mr. Bent had done. Jack gave his quick little chuckle and said:

"It seems you didn't know some of your neighbors very well. Now that he is dead, I guess there is no harm in telling. Bent was not his real name—it was Jenkins. When he was a young man, he ran away from home, joined one of the notorious outlaw gangs in Oklahoma, and stayed with them for several years. In the long periods of inactivity of such a life, he taught himself to read Greek, because a grandfather he much admired, a preacher, could read Greek. He came to see that the life of an outlaw was leading nowhere but into trouble, with no financial gain to justify the risk, so he decided to quit. He went back to his home, in the mountains of Tennessee, married his childhood sweetheart, and changed his name. The initial "A" in his new name was for Aristobolus, a hero in a Greek book he liked. He then tried to get as far away from the old gang and everybody that knew him as he possibly could.

"They traveled west until they came to the ocean and could go no further. His first act was to report to the sheriff (that was before my time), and as long as he lived he formally reported every six months, as is required by Oregon law of everyone living under an assumed name. He got a job, saved his money, and started that little store eventually. When I was elected sheriff, he was the first man I appointed deputy. That old man wasn't afraid of anything that walks, flies, swims, or crawls, and he was a dead shot with a revolver, rifle or shotgun. I would rather have him side me in a tight spot than any other man in Tillamook County."

You can't always, sometimes, tell.

These simple little tales tell nothing earth-shaking; they are just typical extracts from the business of day-by-day living, as seen and heard by a country doctor in the Big Timber Country. I don't think anybody had more fun than he did. It was great to be young and enjoy life. It was great to be young in such a beautiful land, with its forest covered mountains, the streams full of fish, the ocean, the lush pastures with grazing cattle; the hardworking, friendly, helpful people. I guess you could say, "Those were the Good Old Days," though we did not recognize them as such at the time.

Things are different now. Changes in industrial techniques, changes in agricultural techniques, roads, automobiles, improved communications, financial inflation, installment buying, application of scientific advance to medical and other fields, distribution of conveniences and comforts—all have had their effect, for good or ill. Many of the good things of that period have been lost. The old way of life is gone forever, but I hope this little record will help to preserve its memory.

Howard Rigsby, in *The Tulip Tree* (©1963, Doubleday & Company, Inc.), said, "It is our memories, not our arteries, that set us apart, we old ones. It's because we can remember how it was. It's in *our* lifetime, only ours, since the beginning, that *all* has changed. Life couldn't have been much different in George Washington's day than it was in Julius Caesar's. *We* are the ones who have witnessed the birth of automatic man."

I like to think about the old times—that's an old man's privilege. I like to remember the friends, many of whom are no longer among the living, and the good times we had together. I like to recall little incidents like those recorded here, but I am glad they are behind me, not ahead. I wouldn't turn back the calendar if I could.

Hebo, southern Tillamook County, near Mt. Hebo, circa 1920s.
(OHS Collections)

PACIFIC RAILWAY TRAIN NEAR